INTRODUCTION TO HERPETOLOGY

OTHER BOOKS
IN BIOLOGY

The Study of Plant Communities: An Introduction to Plant Ecology (Second Edition)
  Henry J. Oosting

Principles of Human Genetics (Second Edition)
  Curt Stern

Experiments in General Biology
  Graham DuShane and David Regnery

Principles of Plant Physiology
  James Bonner and Arthur W. Galston

General Genetics
  Adrian M. Srb and Ray D. Owen

An Introduction to Bacterial Physiology (Second Edition)
  Evelyn L. Oginsky and Wayne W. Umbreit

Laboratory Studies in Biology: Observations and their Implications
  *Chester A. Lawson, Ralph W. Lewis, Mary Alice Burmester, and Garrett
  Hardin*

Plants in Action: A Laboratory Manual of Plant Physiology
  Leonard Machlis and John G. Torrey

Comparative Morphology of Vascular Plants
  Adriance S. Foster and Ernest M. Gifford, Jr.

Taxonomy of Flowering Plants
  C. L. Porter

Growth, Development, and Pattern
  N. J. Berrill

Biology: Its Principles and Implications
  Garrett Hardin

Genetics and Evolution: Selected Papers of A. H. Sturtevant
  E. B. Lewis, Editor

Animal Tissue Techniques
  Gretchen L. Humason

Microbes in Action: A Laboratory Manual of Microbiology
  Harry W. Seeley, Jr., and Paul J. VanDemark

Botanical Histochemistry: Principles and Practice
  William A. Jensen

Modern Microbiology
  Wayne W. Umbreit

Laboratory Outlines in Biology
  Peter Abramoff and Robert G. Thomson

Molecular Biology of Bacterial Viruses
  Gunther S. Stent

Principles of Numerical Taxonomy
  Robert R. Sokal and Peter H. A. Sneath

Structure and Function in the Nervous Systems of Invertebrates
  Theodore Holmes Bullock and G. Adrian Horridge

Introduction to

HERPETOLOGY

BY COLEMAN J. GOIN

University of Florida

AND OLIVE B. GOIN

W. H. FREEMAN
AND COMPANY

San Francisco and London

PREFACE

This book is planned for use as a text in a one-semester course in herpetology. It is designed for students who have had one year of college biology, but who may have had no more than one year.

Courses in herpetology usually consist of two parts—a series of lectures or discussions in which basic principles are presented, and a series of laboratory and field exercises. We believe that laboratory work should be based primarily on local faunas and on the specimens available in local institutions. These differ from region to region. Techniques useful for collecting animals in one climate may be of little use in another; the season of the year at which animals are abundant and active varies from place to place; the characters used in identifying specimens from one faunal region may not apply to those from another. Moreover, excellent field guides, keys, and local lists are available for most places where courses of herpetology are presented today. We have therefore left the organization of the laboratory and field work to the individual instructor, and have concentrated instead on the major aspects of herpetology that apply throughout the world.

In these days when so much of biology is concerned with happenings at the molecular level within the individual cell, we believe there is a definite need for the student to appreciate that these processes have biological meaning only as they help us to understand the living, functioning animal. We agree with Professor Romer that:

"It is not enough to name an animal; we want to know everything about him: what sort of a life he leads, his habits and instincts, how he gains his food and escapes enemies, his relations to other animals and his physical environment, his courtship and reproduction, care of his young, home life (if any). Some aspects of these inquiries are dignified by such names as *ecology* and *ethology*; for the most part they come broadly under the term *natural history*. Many workers who may study deeply—but narrowly—the physiological processes or anatomical structure of animals are liable to phrase this, somewhat scornfully, as 'mere natural history.' But, on reflection, this attitude is the exact opposite of the proper one. No anatomical structure, however beautifully designed, no physiological or biochemical process, however interesting to the technical worker, is of importance except insofar as it contributes to the survival and welfare of the animal. The

v

study of the functioning of an animal in nature—to put it crudely, how he goes about his business of being an animal—is in many regards the highest possible level of biological investigation." (*The Vertebrate Story,* A. S. Romer, Univ. Chicago Press, Copyright 1959 by the University of Chicago.)

In preparing this volume, our task has been twofold: we have had first to decide on the basic organization that we felt a text in this field should have, and second to synthesize a mountain of original literature into a volume of modest size.

In the former task, we have worked under the firm conviction that the proper approach was to discuss basic biological principles as exemplified by amphibians and reptiles. This we have tried to do. In the latter task, we have of necessity shown a great deal of personal bias in deciding just what should and what should not be included. Because of the wealth of original literature, there are many fascinating facts of herpetology that we have had to leave unmentioned. We know that some of our professional colleagues will feel that, like good Anglicans, "we have left out those things which we ought to have put in, and we have put in those things which we ought to have left out."

In the first chapter we indicate the position of the amphibians and reptiles in the animal kingdom, discuss briefly certain basic principles of classification, and give a résumé of the rise of herpetology as a science. The next four chapters deal with the structure and evolutionary history of the two classes. Chapters 6 through 11 are concerned with natural history and with the mechanisms of speciation and geographic distribution. The last six chapters give a summary of the living amphibians and reptiles to the family or subfamily level, with notes on life history and geographic distribution.

Since herpetology is a worldwide subject, we have tried, in our choice of examples and illustrative material, to strike a balance between native and exotic forms.

The references given at the end of each chapter are intended simply as suggestions to the interested student who may wish to pursue a particular topic further than is possible in an introductory text. Most are compendiums. Occasionally, we have included original papers that are of exceptional interest and importance and that give information not generally available elsewhere.

It is a pleasure to acknowledge the help we have received from so many people in the preparation of this text. We are indebted to M. Graham Netting for reading Chapters 8, 9, 11, 13, and 17, and to Archie Carr for reading Chapters 8, 9, and 14. We wish especially to thank Kenneth W.

Cooper, not only for the critical reading of Chapter 10, but also for the continuing interest he has shown in this work, and for the many pertinent references he has sent us. Henryk Szarski read Chapter 4 and Carl Gans read portions of an earlier draft of the manuscript. Walter Auffenberg assisted us materially in dealing with the classification of the snakes and crocodilians.

Mr. and Mrs. J. C. Battersby, Charles M. Bogert, Robert F. Inger, and Alfred S. Romer all responded most kindly to appeals for special information and material.

Not least have been the intangible benefits we have received from discussions with these and other colleagues, among whom we should like to mention in particular Doris M. Cochran and Ivor Griffiths.

Except where otherwise indicated, the excellent photographs were made by Isabelle Hunt Conant, many of them especially for this volume. The line drawings are from the gifted pen of Evan Gillespie, many of them from sketches made originally by Esther Coogle.

To James E. Böhlke, Alice G. C. Grandison, W. S. Pitt, Oswaldo Reig, and Charles K. Weichert we are indebted for special illustrative material.

We wish to thank Dean George T. Harrell for the use of a dictaphone during the preparation of the first draft, and Mrs. Sue P. Johnson for her careful typing of the final draft.

Since everything we have ever done has had imperfections, we feel sure that this book will have its share. We would like to request that our friends be kind enough to point out to us our errors, both of omission and of commission, so that in the future we may mend our ways.

Coleman J. Goin

March 1962   Olive B. Goin

CONTENTS

INTRODUCTION

THE SCIENCE of biology has become enormously complex. It is no longer possible for one person to encompass all of its ramifications. To comprehend the subject matter, workers in the field have been increasingly obliged to limit themselves to various subsciences, such as anatomy, genetics, embryology, or ecology. These subsciences can be visualized as extending vertically through the parent science. But this is not the only way biology can be approached; it can be subdivided according to the kinds of organisms studied. These subdivisions include such disciplines as ornithology, entomology, and herpetology. They extend across and interweave with the primary divisions mentioned above.

Biology might thus be visualized as a vast tapestry with the threads of the warp formed of the many subsciences and those of the woof formed of the groups of organisms under study. Whether one wishes to follow a thread of the warp and study something like the anatomy of one or more structures in many different kinds of animals (comparative anatomy), or to follow a thread of the woof and study the anatomy, behavior, and distribution of a particular group of organisms like snakes, is a matter of personal inclination. When we study herpetology (*herpeton* = crawling thing, *logos* = reason or knowledge) we follow those threads of the woof that are made up of the amphibians and reptiles. Any aspect of the biology of these animals is legitimately a part of the subject matter of herpetology.

ZOOLOGICAL POSITION OF AMPHIBIANS AND REPTILES

We cannot profitably study any of the broader aspects of zoology until we know just what animals we are dealing with and where they fit in the whole pattern. No one knows exactly how many different kinds of

1

animals there are, but a recent careful estimate gives 1,120,000 and this is probably not too far from actuality. (To bring some order and meaning into this bewildering array of forms, man has found it necessary to classify and divide them into groups and categories. There are a number of different schemes of classification we might adopt. We might, for example, divide animals according to the habitats in which they live, such as forest, desert, lake, or ocean. Sometimes, for certain types of studies, this classification is used. But the standard, universally accepted classification of today, the one we mean when we speak of animal classification, is based on degree of relationship through descent from a common ancestor, usually judged by similarity of morphological characters.)

The animal kingdom is divided into a number of great groups called phyla, such as the phylum Mollusca (shellfishes, like the oysters, clams, snails, and squids) or the phylum Arthropoda (joint-footed animals, like the insects, spiders, centipedes, and lobsters). The phylum Chordata comprises those animals that at some stage in their life history have gill slits, a hollow, dorsal nerve cord, and a notochord (a stiffening rod running along the back). The phylum Chordata is divided into several small and unimportant subphyla and one large one, the subphylum Vertebrata, to which the animals most familiar to us, the fishes, amphibians, reptiles, birds, and mammals, belong. Vertebrates are animals in which the notochord, though still present in the early embryonic stages, has been largely replaced in the adults by a jointed vertebral column composed of a number of separate structures, the vertebrae. The anterior end of the nerve cord is expanded to form a brain which is enclosed in a protective box, the cranium.

The members of the subphylum Vertebrata are divided into a number of classes. There are several classes of fishlike vertebrates plus the classes Amphibia, Reptilia, Aves, and Mammalia. These include about 40,000 different kinds of living animals:

Fishlike vertebrates	20,000±	species
Amphibians	2,100±	species
Reptiles	6,000±	species
Birds	8,600±	species
Mammals	3,200±	species

Amphibia

The amphibians are backboned animals whose body temperature is under the control of the external environment; that is, they are ectothermal. They have soft glandular skins, for the most part without scales. They lack the paired fins of the fishes; instead most have limbs with digits as do the higher forms—the reptiles, birds, and mammals. The four classes of animals with limbs are sometimes linked in the superclass Tetrapoda (four-footed). The amphibian egg lacks a shell; to keep the developing embryo from desiccating, the egg must be laid in water or in humid surroundings.

The ancestral amphibians were derived from primitive fish. They were the first vertebrates to move out on land and they gave rise to all the other terrestrial vertebrates, the reptiles, birds, and mammals. Two hundred fifty million years ago they were a prominent element in the world's fauna, but today they are a decadent stock with only about two thousand living members. These are divided into four orders:

Order Apoda (Caecilians)	75± species
Order Trachystomata (Sirens)	3 species
Order Caudata (Salamanders)	280± species
Order Anura (Frogs and Toads)	1,800± species

Reptilia

The reptiles, like the amphibians, are ectothermal vertebrates without paired fins. They differ from the amphibians in that they all have scales. The reptile egg usually has a calcareous shell and is laid on land. A new membrane, the amnion, forms a fluid-filled sac in which the embryo develops. Together with the shell, this sac protects the embryo from desiccation. Since an amnion also surrounds the embryos of birds and mammals, these three classes are sometimes called amniotes to distinguish them from the anamniotes, the fishes and amphibians.

Descendants of the early amphibians, the reptiles were the dominant animals of the earth during the Mesozoic era and gave rise to the two classes of vertebrate animals that have internal temperature control, the endothermal birds and mammals. The reptiles have lost the dominant position they held during the Mesozoic, although they are still much more numerous than the amphibians. The living forms are divided into the following orders:

Order Testudinata (Turtles)	335	species
Order Rhynchocephalia (Tuatara)	1	species
Order Squamata {Suborder Lacertilia[1] (Lizards)	3,000±	species
{Suborder Serpentes (Snakes)	2,700±	species
Order Crocodilia (Alligators and Crocodiles)	21	species

[1] The Greek forms, Sauria and Ophidia are often used for the suborders of lizards and snakes, respectively, but we prefer the Latin forms, Lacertilia and Serpentes, wh¹ch agree with the ordinal name, Squamata.

The amphibians and reptiles have traditionally been studied together. This is partly for historical reasons, since originally the differences between the two groups were not recognized as being important enough to justify their placement in separate classes. Partly it is a matter of convenience. The amphibians are a small group and the methods of collecting and preserving them are similar to those used for reptiles. To avoid repeating the rather cumbersome phrase "amphibians and reptiles" we need a general term for the members of these two classes. "Herptiles" has been proposed and will be used as a synonym for that phrase in this book.

With knowledge becoming more and more detailed and the literature more and more voluminous, it is difficult for a worker even in the restricted field of herpetology to keep abreast of current developments. Some modern herpetologists restrict themselves almost entirely to the study of the amphibians, or of the reptiles, or perhaps even of a single order, such as the Testudinata.

SYSTEMATICS AND TAXONOMY

The task of naming, describing, and classifying the amphibians and reptiles has necessarily had priority, and even at the present time much of the literature of herpetology is concerned with such studies. Here herpetology interweaves with taxonomy and systematics. These two terms are often used interchangeably, but it seems better to restrict taxonomy to the frequently very complicated task of assigning names to groups of animals. Systematics is concerned with working out a classification that will be descriptive of the relationships of the animals. The two do overlap, of course. Current taxonomic practice requires that the taxonomist describe the form he is naming and include in the description an indication of the relationships of the animal. The systematist frequently finds that he must name one or more new forms or straighten out a nomenclatural problem before he can discuss intelligibly the relationships of the animals he is classifying.

Our present system of classification comprises a series of categories, each less inclusive than the preceding one. Phyla are divided into classes, classes into orders, orders into families, families into genera (sing. genus) and genera into species (sing. species). A Leopard Frog is classified thus:

Phylum Chordata
Class Amphibia
Order Anura
Family Ranidae
Genus *Rana*
Species *Rana pipiens*

These are the standard categories, but as classification has grown more precise we have come to feel the need for additional groupings. So we sometimes have superclasses,[1] infraorders, subfamilies and so forth. If a genus includes a large number of rather diverse species which fall into natural groups, it may be divided into several subgenera. If it includes only a few quite similar species, subgenera are not necessary. Every animal is a member of a species, a genus, a family, an order, a class, and a phylum; it may or may not belong to a supergenus or suborder.

The basic systematic unit is the species. Species are generally what we have in mind when we talk about "kinds" of organisms. Yet simple as this seems on the surface, probably no concept in all biology is more argued about at the present time than the species concept. Biologists do not even agree as to whether a species has objective reality or is subjective, a convenient pigeonhole devised by man to simplify the handling of data. Most of them do agree, though, that the species is somehow different from the other categories.

There have been innumerable attempts to define the word "species," but none of them are completely satisfactory for all organisms. Considering the multiplicity of animal and plant forms, and the different ways they have of reproducing themselves, this is not surprising. Any definition that covered all would either be so vague as to be meaningless or would expand into an essay or even a book and would thus no longer be a definition. Fortunately, we are here dealing only with amphibians and reptiles, animals that are multicellular, bisexual, and not self-fertilizing. For such animals it is possible to frame a working definition. *A species is a population of animals that freely interbreed in nature (or would do so if brought together) to produce fertile offspring.* The concept of a particular species may indeed be a creation of the mind of man, but the population on which it is based is real. The conscientious systematist tries to frame and express his concept so that it will conform as closely as possible to that objective reality.

When a population is spread over a geographic area comprising diverse environmental conditions, the animals in one part of the range frequently differ slightly from those in another part. These different parts of the population are called geographic races or subspecies. They are not reproductively isolated, since in areas where two races meet they do interbreed. The part of the population in such an area is intermediate in character between the two races. These are called areas of intergradation.

The criterion of whether two forms can and do interbreed is basic to the

[1] The prefixes (super = above, infra and sub = below) indicate where the category falls in relation to the standard ones given above. Thus a superclass is a grouping of classes within a subphylum or phylum.

modern species concept, but unfortunately it is often very difficult to apply
in actual practice. Except for the frogs, which proclaim their intentions to
high heaven, we are seldom able to observe breeding in the field. Nor does
it always help to bring the animals into the laboratory. Sometimes two
forms that do not ordinarily cross in nature will do so under artificial conditions; other animals refuse to mate at all in captivity.

Consider the plight of a biologist who has received two collections of
reptiles from two different localities in Africa. In one is a series of snakes, all
much alike and different from other specimens in the collection. In the
second collection is a series of similar snakes that are slightly smaller, have
yellow stripes on a black background instead of tannish stripes on a dark
brown background, and have an average of 150 ventral scales rather than
170. Do these two forms interbreed in nature? Are they two species or two
geographic races of one species? Short of organizing an expedition to Africa
and hunting for an area of intergradation between the two, how is our
biologist to find out? Must he refrain from publishing on these collections
and describing and naming the snakes until he does find out? Fortunately,
no, otherwise we should never get on with the business of naming and
classifying the animals of the world. We know from experience that, in the
amphibians and reptiles at least, two forms that are reproductively isolated
usually differ in morphological characters and that *in general* the morphological differences between two species are greater than the ones between
two subspecies. Here the experience and judgment of the biologist come
into play. He simply decides whether the two forms are near enough alike
so that they probably do interbreed, and describes them as subspecies or
species accordingly. It should be stressed, though, that degree of morphological difference is an indication of, but does not determine, degree of
relationship. Sometimes two closely similar forms are reproductively isolated, hence are distinct species.

Morphological differences alone do not make a species, but they *are*
correlated with other differences, in physiology, cytology, behavior, and
ecological requirements, that result in reproductive isolation and do, in sum
total, make a species. We use the morphological characters most often because they are the ones we can see, measure, and compare in the preserved
material with which the systematist usually must work.

Furthermore, we must always remember that species are products of
evolution and that evolution is a gradual process. It would be surprising
indeed if we did not sometimes find two populations that are still in the
process of becoming species, that have not quite achieved complete reproductive isolation from each other. Here again the systematist must use
his own judgment in deciding whether or not the degree of isolation is

sufficient to warrant calling the two forms species. The decisions of the systematist are by no means final. Collection of further data often shows that two forms originally described as separate species are really geographic races of one species and sometimes what was called a subspecies is found to be a full species. Fortunately, under our current system of nomenclature, it is easy to make such shifts in rank without upsetting the whole scheme of classification.

The systematist must also distinguish between morphological similarities due to relationship and those that sometimes develop through parallel or convergent evolution in forms really quite distantly related. When two separate lines of organisms become adapted to the same basic type of environment in different regions of the world, their members may come to resemble each other closely. This has happened frequently among the herptiles. Several montane species of Asiatic land salamanders of the family Hynobiidae have reduced lungs and one genus, *Onychodactylus,* has lost the lungs entirely, a loss that is characteristic of the distantly related mountain salamanders of the family Plethodontidae. Some of the Old World tree frogs of the genus *Rhacophorus* (family Rhacophoridae) look so much like some South American tree frogs of the genus *Phyllomedusa* (family Hylidae) that it takes close inspection to distinguish between them. Among the reptiles, the Moloch (*Moloch horridus*) of Australia, a lizard of the family Agamidae, has a habitus very similar to that of the horned lizards (*Phrynosoma*) of western North America, members of the family Iguanidae. Both live in arid regions and feed almost exclusively on ants. There is, of course, the classic case of the Sidewinder, *Crotalus cerastes,* a rattlesnake of the deserts of western United States, that resembles so closely the Sand Viper (*Cerastes*) of the deserts of northern Africa and southwestern Asia that its specific name was given in recognition of this similarity. The likeness is further emphasized by their similar sidewinding locomotion.

NOMENCLATURE

Our system of nomenclature is based on the one first universally applied to all animals by Linnaeus in the tenth edition of his *Systema Naturae,* published in 1758. Linnaeus gave each species known to him a name consisting of two parts, the name of the genus to which it belonged, plus a specific epithet, the trivial name (e.g., *Rana esculenta* Linnaeus). Over the years other biologists adopted the Linnaean system of designating species. But as more and more new species were found and named, and more and more papers were published using these names, confusion inevitably crept in. Sometimes a man, either deliberately or through ignorance, gave a name to

a species that had already been named something else, thereby creating a
synonym. Sometimes two men, working on different groups, happened to
give the same name to entirely different forms. Such a name is called a
homonym. But if we, as biologists, are to understand one another, we must
be sure that each species has only one name and that each name applies
to only one species.

The need for a set of rules for taxonomists to follow in proposing new
names, or in deciding which of several names already in use should be accepted, soon became acute. A number of codes were drawn up in different
countries and for different groups. The English followed one code, the
French another, the Germans still a third, the ornithologists had a code of
their own, and so did the paleontologists. Finally, in 1895, the Third International Zoological Congress appointed a Committee which drew up the
International Rules of Zoological Nomenclature, commonly known as the
Code. This Code was accepted by the Fifth International Zoological Congress in 1901 and is now universally followed. It has been revised from time
to time, most recently in 1961. A permanent International Commission of
Zoological Nomenclature serves as a Supreme Court to resolve the knotty
problems of interpretation that seem to be inevitable under any code of
laws.

Most of the rules and recommendations of the Code deal with technical-
ities of interest only to professional taxonomists. A few of them should be
familiar to every zoologist, if only because an understanding of them will
clarify his reading and facilitate his writing on zoological subjects.

No genus may have the same name as any other genus of animals, but the
same trivial name may be used in different genera (though not more than
once within a single genus). The specific name of an animal consists of the
generic name plus the trivial name. Thus the specific name of the Leopard
Frog classified in the preceding section is not *pipiens* but *Rana pipiens*.
When a specific name is written, the generic name always begins with a
capital letter, the trivial name with a small letter, and both are italicized,
indicated in a manuscript by underlining the name. If a species is found to
comprise two or more geographic races, each is given a third name which
follows the trivial name and like it is never capitalized. The third name of the
subspecies that most nearly represents the material on which the trivial
name of the species was based is always the same as the trivial name. Thus
the race of the Leopard Frog in eastern United States is *Rana pipiens
pipiens* while the Florida race is *R. p. sphenocephala*.[1] Sometimes a sci-

---

[1] Note that when the full name has been written out once in a paragraph, we may, if
we use the name again in the same paragraph, abbreviate the generic name and, when
a trinomial is used, the trivial name.

entific name is followed by the name of the man who first proposed it, e.g., *R. p. pipiens* Schreber. The name of a higher category (family, order, class, *etc.*) begins with a capital letter but is not italicized in print. Family names always end in -idae (Ranidae) and subfamily names in -inae (Raninae).

HISTORY OF HERPETOLOGY

Among the early writers on natural history, the names of Aristotle and Pliny stand out. Aristotle (384–322 B.C.) has been called the originator of biological classification, not because his system bears much resemblance to the one we use today, but because he first felt the need to establish categories based on characteristics, particularly anatomical ones, of the animals. He says "Four-footed beasts that produce their young alive have hair; four-footed beasts that lay eggs have scales." Compare this with the Biblical division of animals as "clean" or "unclean."

The Roman Pliny (A.D. 23–79) wrote a *Naturalis Historiae* in which he listed all the animals he knew or had heard of. Pliny was no systematist and there is little order to his arrangement. He was also very credulous. Many of his animals are fabulous and many of the tales he tells of real animals are equally fabulous. But his accounts are lively and, where he deals with animals he could study himself, they do contain some accurate information.

The works of these two men were handed down in manuscript form during the Middle Ages, and were among the first books made widely available after the invention of printing in the mid-fifteenth century. Pliny's Natural History was published in 1469 and Aristotle's *Editio Princeps*, containing his *Historia de Animalium*, was printed in the period 1495–1498.

The reawakened interest in the world of nature that came with the Renaissance, and the enormous widening of the horizons of that world following the voyages of the explorers, resulted in the discovery of many more kinds of animals. The works of Pliny and Aristotle no longer sufficed. Several new compendia appeared, largely based on the classification of Aristotle and modeled on the style of Pliny. Notable is the *Historia Animalium*, published between 1559 and 1583 by Conradus von Gesner, a talented Swiss naturalist. Gesner's greatest contribution was the introduction of scientific illustrations. His woodcuts of a frog, a toad, a ringed snake, and a viper are probably the first published figures of amphibians and reptiles.

The first book in the English language on natural history is that of the Reverend Edward Topsell who, in 1608, published a volume entitled *Historie of Foure-footed Beasts, describing the true and lively figures of every Beast. . . .collected out of all the volumes of C. Gesner and other*

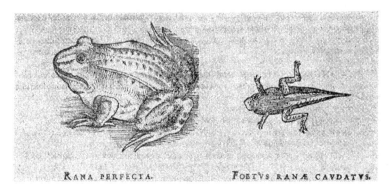

RANA PERFECTA.   FOETVS RANÆ CAVDATVS.

FIG. 1-1. "Rana perfecta" as illustrated by Gesner in his *Historia Animalium,*
1586, p. 46, one of the earliest herpetological illustrations. The name is not a
specific name in the Linnaean sense, but means the adult or "perfect" form of
the frog. [Courtesy the United States National Museum.]

writers. The next year Topsell brought out his *Historie of Serpents or the
Seconde Book of living creatures.* Topsell's conception of serpents included
practically everything that creeps and crawls and hence involved spiders,
scorpions, and many fabulous creatures, in addition to the amphibians and
reptiles.

The first truly critical and systematic approach to herpetology was the
work of John Ray, an Englishman, son of a village blacksmith. He was
primarily a botanist but his *Synopsis Methodica Animalium Quadrupedum
et Serpentini Generis,* published in 1693, is concerned with the herptiles.
Ray was the first to group these animals together for the reason that their
hearts have a single ventricle in contrast to the two-chambered ventricle of
the birds and mammals. He further distinguished between the harmless
and poisonous snakes by the character of their teeth. His work was really
excellent for the time.

It is unfortunate that Linnaeus (1707–1788) did not adopt Ray's classification. He named the herptiles he knew and gave them a place in his
classification, but he had little interest in or liking for them, as is shown
by his summary in *Systema Naturae:*

"These foul and loathsome animals are distinguished by a heart with a
single ventricle and a single auricle, doubtful lungs and a double penis.

"Most amphibia are abhorrent because of their cold body, pale colour,
cartilaginous skeleton, filthy skin, fierce aspect, calculating eye, offensive
smell, harsh voice, squalid habitation, and terrible venom; and so their
Creator has not exerted his powers (to make) many of them."

·The classification of Linnaeus is faulty and inadequate, but the principles he laid down and the system of nomenclature he established allowed other workers to place systematic herpetology on a firm foundation.

In post-Linnaean times, herpetology has made rapid advances. We shall here mention only a few of the more important early workers. Among these is André Marie Constant Duméril, a French naturalist and herpetologist. With his colleague, Gabriel Bibron, Duméril wrote an important ten-volume work, *Erpétologie Générale*, which was published from 1835 to 1854. Some of these volumes were co-authored by Auguste Henri André Duméril.

Albert C. D. G. Günther, who served as Keeper of Reptiles in the British Museum from 1856 to 1895, gave tremendous impetus to herpetology when he published his series of catalogues of amphibians and reptiles in the British Museum. He also did spadework for Neotropical herpetology in his volume on amphibians and reptiles in the series *Biologia Centrale Americana*. George Albert Boulenger, who worked in the British Museum from 1882 to 1920, laid the groundwork for worldwide herpetology with a nine-volume catalogue of the batrachians and reptiles in the British Museum.

In the New World, Edward Drinker Cope (1840–1897) was publishing paper after paper on fundamental herpetology. Cope not only made major

FIG. 1-2.
George Albert Boulenger. [Courtesy the British Museum (Natural History).]

contributions in a tremendous number of descriptive papers on amphibians and reptiles, but also, through his great knowledge of comparative anatomy and paleontology, established a foundation for much of our modern classification. Two of his most important works were published after his death. *The Batrachia of North America* (1889), and *The Crocodilians, Lizards, and Snakes of North America* (1900), are still fundamental books in the library of anyone concerned with North American herpetology.

The name of Leonard Stejneger (1851–1943) is perhaps not appreciated as much as it should be by many younger North American herpetologists. From the time he was appointed Curator of Reptiles in the United States

National Museum in 1889, until his death in 1943, he published volumi-
nously on the herpetology, not only of America but of the world. Among his
important books are the *Herpetology of Japan* and *The Poisonous Reptiles
of North America*. But perhaps his major contribution was a work he under-
took with his younger friend and associate, Thomas Barbour, Director of the
Museum of Comparative Zoölogy. In 1917 the two brought out the first
edition of the *Check List of North American Amphibians and Reptiles*, a
volume that went through five successive editions under their joint author-
ship. It is hard indeed for many young herpetologists to appreciate fully the
difficulties faced by older workers in the
days before a checklist and handbooks
for every group were readily available.
Stejneger and Barbour, by synthesizing
North American herpetology into a
readily digestible form, had an incal-
culable effect on stimulating research
in the field.

FIG. 1-3.
Edward Drinker Cope. [Courtesy
the Academy of Natural Sciences
of Philadelphia.]

Finally we believe it only fair to mention the name of Raymond L. Ditmars,
late Curator of Reptiles at the New York
Zoological Society. At a time when the
publication of popular books was
frowned upon as unworthy of a true
scientist, Ditmars persisted and turned
out a series of popular volumes on reptiles, particularly snakes, that awakened
many a youngster to an interest in herpetology. It was only after this interest
was developed that these young workers
could begin to digest the more fundamental work of such men as Blanchard,
Dunn, Noble, Schmidt, and Van Den-
burgh.

Actually, Ditmars was performing one of the most important functions of
a scientist; one that is all too often overlooked. A scientist does have the
responsibility for interpreting his speciality to the general public when
called on to do so and for answering the questions of amateurs and interested laymen. Since Ditmars' time, several other herpetologists have recognized this responsibility and have written excellent popular books on the
herptiles.

These books are helping to diminish the widespread, long-standing, and unwarranted public antipathy toward the herptiles. Nevertheless, too many people still share the opinion of Linnaeus and wonder why anyone should waste his time on such unpleasant creatures. One answer is that a thorough knowledge of the herptiles will elucidate most of the principles that apply throughout the animal kingdom and may provide a clue to some of the still-unanswered problems of basic biology. But in the last analysis, the herpetologist studies reptiles and amphibians because, in contrast to Linnaeus, he likes them and finds them interesting. *De gustibus non est disputandum.*

Collateral Reading and General Reference

Hemming, Francis, ed. *Copenhagen Decisions on Zoological Nomenclature.* London: International Trust for Zoological Nomenclature, 1953.

International Code of Zoological Nomenclature. London: International Trust for Zoological Nomenclature, 1961.

Mayr, Ernst, E. G. Linsley, and R. L. Usinger. *Methods and Principles of Systematic Zoology.* New York: McGraw-Hill, 1953. (A stimulating discussion of the modern approach to problems of systematics and taxonomy.)

Nordenskiöld, Erik. *History of Biology.* New York: Tudor, 1928. (Perhaps the best single book on the history of biology; excellent accounts of the lives and works of great biologists, and the relationships of each to the historical events and intellectual climate of his time.)

Romer, A. S. *The Vertebrate Story.* Chicago: University of Chicago Press, 1959. (A very readable, well-illustrated account of the evolution and natural history of the vertebrates.)

STRUCTURE OF
AMPHIBIA

WITHOUT THE TRANSITION made by the amphibians from an aquatic to a terrestrial existence the evolution of the higher vertebrates could not have taken place. The cause and course of this major shift have been subjects of much speculation. The individual modern amphibian accomplishes the same transition in weeks or months, instead of thousands of years, and because we are so familiar with it we accept it as one of the commonplaces of nature. However, for neither the individual nor the group is the transition complete; most adult amphibians are bound to humid environments and all of them that lay eggs must do so either in water or in moist surroundings. Their transitional position imposes structural and functional complications. Some of their characteristics are carry-overs from fish ancestors, others look forward to the more terrestrial reptiles to come. Some are adaptations to life in the water, others to life on land. Some reflect the necessities of an individual life cycle that involves a change from an aquatic larva to a terrestrial adult that nevertheless must return to the water to breed. The larva, moreover, does not resemble the fish ancestor of the amphibians, and the steps by which it metamorphoses into an adult are not the same as the steps by which the fish evolved into an amphibian. The tadpole, for instance, does not start out with fins that change into legs.

The amphibians have no unique structures, such as the feathers of birds or the mammary glands of mammals, that set them off from the other groups of tetrapods. There is no structure of which we can say, "This is found in all amphibians and only in amphibians." Even the characteristic life history which gives the group its name (*amphibios* = double life) does not hold

14

for all members, since some frogs develop directly from the egg into tiny frogs without going through a tadpole stage, and some salamanders bear their young alive. Yet no one doubts that these animals are true amphibians. Therefore, any definition of the class Amphibia must be based on a combination of characters rather than on definitive, unique ones.

DEFINITION OF AMPHIBIA

Members of the class Amphibia—the frogs, salamanders, sirens, and caecilians—are usually small animals, having smooth, moist skins without scales. They lay their eggs in water or in moist surroundings. The egg is covered by several gelatinous envelopes rather than a shell and usually hatches into a larva. The larva differs structurally from the adult and metamorphoses, that is, changes abruptly rather than grows gradually, into the adult body form.

The larvae have gills which may be enclosed within a gill chamber. In the adults, respiration takes place through lungs, through gills, directly through the skin or mucous membrane lining the mouth and pharynx, or by means of some combination of these. The lungs are simple in structure and usually appear at metamorphosis.

Amphibians are ectothermic. Multicellular mucous and poison glands occur in the skin. There are no true nails or claws, although in some forms horny epidermal structures may be present at the tips of the toes. The heart is three-chambered. The skull is flattened and composed of fewer bones than is the fish skull; it articulates with the vertebral column by two occipital condyles (a condition characteristic of mammals but not of reptiles or birds). As in the fish, there are only ten pairs of cranial nerves, rather than the twelve pairs present in the higher tetrapods.

ANATOMY OF AMPHIBIA

Detailed anatomical descriptions will not be presented in this book. We are here concerned chiefly with those structural advances from the condition in fishes which, in combination, make an animal an amphibian.

Integument

We are accustomed to think of the evolution from the aquatic fish to the more or less terrestrial amphibian as involving mainly the shift from aquatic to aerial respiration and the transformation of fins into legs. Actually, the process called for basic structural changes throughout the body. Some of the most important of these took place in the integument, since the skin is after all the part directly in contact with the external environment.

The amphibian skin, as in all vertebrates, is made up of two parts, an outer epidermis and an inner dermis.

Epidermis. The epidermis comprises several layers of cells, including a stratum corneum, an outer layer of horny, dead cells that protects the deeper living cells and helps prevent moisture loss. This adaptation to life on land, which is not present in fish, is absent in only a few aquatic salamanders among the amphibians. During ecdysis, the shedding of the skin, it is the stratum corneum that is sloughed off. In some forms (e.g., the Leopard Frog, *Rana pipiens*) the skin comes off in bits and pieces. In others (e.g., the toads, *Bufo*) it is pulled off as a whole. We never find a shed toad skin as we do a snake skin because the toad eats it immediately. The epidermis of the larva is made up of two layers from the initial stage.

Dermis. The dermis is relatively thin and does not differ basically from that of fish. Since it often serves as a respiratory organ, it is usually well supplied with blood vessels. Pigment in the amphibians is found in special cells called chromatophores usually located in the uppermost layer of the dermis or between the dermis and the epidermis.

Glands. Epidermal glands may be either unicellular or multicellular. The unicellular glands, so common in fishes, have almost disappeared in amphibians. Patches of them develop in the head region of the embryo and secrete an enzyme that digests the gelatinous envelopes of the egg and so aids in hatching. Unicellular glands known as the glands of Leydig are present in the epidermis of some larval salamanders. Their function is unknown. On the other hand, multicellular mucous and poison glands are numerous and well developed in amphibians. The mucous secretion helps keep the skin moist and provides a medium for the exchange of gases, very important for animals depending wholly or partly on integumentary respiration.

Poison glands are well developed in many different species, especially in the more terrestrial anurans. The so-called warts of toads and the parotoid glands on the back of the neck are actually masses of poison glands. *Hyla vasta*, a giant tree frog found in Haiti, gives off a poisonous secretion so strong that the unwary collector who picks one up with bare hands may suffer inflammation of the skin. Many herpetologists have found all of the frogs in their collecting bags dead at the end of the day except for one species. Apparently the poison of one species may be lethal to members of other species but usually not to individuals of its own kind. Of course the story of getting warts from handling toads is fictitious.

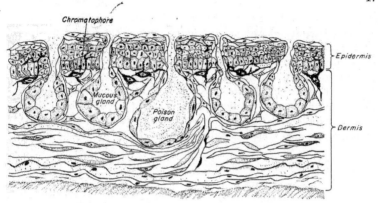

FIG. 2-1. Cross section through the skin of *Rana pipiens.*

Some specially modified integumentary glands are present in the frogs and toads. Many of the tree frogs have glandular discs at the tips of the fingers and toes which apparently aid in climbing, and the glandular thumb pads of breeding male anurans help them to clasp the females.

Scales. As a general rule, the dermal scales characteristic of the fishes have been lost in modern amphibians and the epidermal scales of the reptiles have not yet developed. Rudimentary dermal scales are found buried in the skin of caecilians and a few toads have bony plates imbedded in the skin on the back. The spadefoot toads (*Scaphiopus* and *Pelobates*) have highly cornified areas on the feet which might be considered epidermal scales. The South African Clawed Frog, *Xenopus laevis,* has dark, cornified, epidermal structures, the so-called claws, on the first three digits of the hind foot, and some salamanders, such as *Hynobius,* have cornified, epidermal, clawlike structures at the ends of the digits. These are not true claws, for they lack the underpart, the so-called sole horn or subunguis, of the true claw. They are instead modified epidermal scales.

Digestive System

Tongue. A well-developed, definitive tongue is apparently an adaptation to life on land. Among vertebrates it first appears in the amphibians and is characteristic of all the higher land vertebrates. Food of fishes is already wet when captured and may be swallowed whole. It does not need to be moved around in the mouth for moistening or chewing before it is swallowed. The tongue is a primary tongue, a fleshy fold on the floor of the mouth which lacks intrinsic muscles and can be moved only within narrow limits. It is useful, perhaps, in pushing food farther back in the mouth for swallowing but it cannot be used to capture prey. Some aquatic salamanders (e.g., the water dog *Necturus*) have only a primary tongue. The tongue of some aquatic toads has degenerated and is almost or entirely absent. Other frogs and salamanders have a definitive tongue, which, in addition to the part representing the primary tongue, has an expanded glandular portion supplied with intrinsic musculature. The tongue of most frogs is attached by its base near the anterior margin of the jaw and at rest is folded back on the floor of the mouth with the tip pointing toward the throat. Anyone who has seen a toad snap at a fly knows how quickly the tongue can be flipped out and how accurately it can be aimed. The tongue of some of the salamanders has a more extensive attachment, is mushroom shaped (boletoid tongue), and can be shot forward to trap any unwary insect. In both groups mucous glands in the mouth provide a sticky secretion that clings to the tongue and aids in the capture of prey. Oral glands, like the definitive tongue, are a characteristic of the amphibians not present in lower forms.

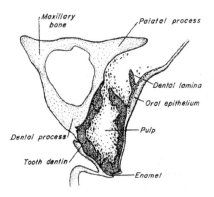

FIG. 2-2.

Section through a maxillary tooth of *Hyla cinerea*, showing its relation to the maxillary bone, oral epithelium, and dental lamina from which the new tooth is formed. [After Goin and Hester, 1962.]

Teeth. With the exception of the epidermal, toothlike structures found in certain larval forms, the teeth of amphibians, as of most other vertebrates, are truth teeth; that is, they have a hard layer of enamel surrounding the softer dentine and central pulp cavity. Toads of the genera *Bufo* and *Pipa* are toothless, but most other anurans and the salamanders have simple, cone-shaped teeth. Except for the genus *Amphignathodon* (*amphi* = both, *gnath* = jaw, *odon* = tooth), frogs have no teeth on the lower jaw. For the most part, amphibian teeth are located on the jaws, but they may also occur on the bones of the roof of the mouth—the palatines and vomers. In a few genera of frogs and plethodontid salamanders they are also attached to the parasphenoid bones. Amphibian teeth are polyphyodont (may be replaced an indefinite number of times), homodont (all the teeth along the jaw are similar), and acrodont (attached to the upper margin of the jawbone instead of to the side).

Gut. The amphibian esophagus is very short, being little more than a constricted area of the alimentary tract. Usually, both the esophagus and the mouth are lined with cilia which sweep small food particles into the stomach. Secreting cells in the esophageal epithelium of some frogs produce an enzyme, pepsin, which does not begin to function until it reaches the stomach. The stomach of some salamanders is simply a straight portion of the gut, whereas in the frogs it has differentiated into a cardiac end leading from the esophagus and a short, narrow, pyloric end leading into the intestine. Digestive glands are absent in the stomachs of some fishes, but are present in all amphibians. The intestine of the caecilians has not yet differentiated into large and small intestines and shows only a slight degree of coiling. Intestinal coiling is more evident in the salamanders, and for the first time there is differentiation into a large and small intestine. In the anurans, the coiling tendency is even more marked, and the large intestine is plainly set off from the small intestine. The amphibian intestine opens into a cloaca, a common chamber into which the urinary, reproductive, and digestive systems empty. A ventral outfolding of the amphibian cloaca gives rise to the urinary bladder.

Respiratory System

By means of the respiratory system oxygen passes from some outside medium, air or water, into the bloodstream of an animal and carbon dioxide passes from the bloodstream into the outside medium. (In a sense all respiration is through water, since respiratory surfaces must be kept constantly moist.) This requires some structure or structures to bring a rich supply of blood into close contact with the medium. Nowhere is the transitional position of the amphibians more clearly shown than in the variety of their respiratory processes. They have three types of highly vascular structures (ones well supplied with blood vessels) that may be used for respiration: gills, lungs, and simply the surface of the skin and the lining of the mouth and pharynx. As a general rule, larval amphibians use gills whereas the adults use lungs, but there are many exceptions. Cutaneous respiration is important in aquatic larvae and in those adults that have moist skins.

Gills. Most fishes have internal gills, which develop from tissue on the inner surfaces of the gill arches, but amphibians have external gills originating from the integument and borne on the outer surfaces of the gill arches. During embryonic development, paired pouches form in the wall of the pharynx. In most amphibians, three of these pairs break through to the outside to form gill slits. Cartilaginous supporting rods, the visceral arches, are formed in the septa between the gill slits. The gills, which consist of filaments covered with ciliated epithelium, are attached to the three visceral arches anterior to the gill slits. Gills of tadpoles are usually smaller and simpler than those of salamander larvae. One caecilian, *Caecilia compressicauda,* has gills that form large, leaflike folds.

Among the frogs, a gill cover, the operculum, develops shortly after the external gills appear. A sheet of tissue grows backward from the hyoid region to cover the gill slits, the gills, and the region from which the forelimbs will eventually develop. It then fuses with the body behind and below the gill region to enclose the gills in an atrial chamber. This chamber has an opening to the outside—the spiracle—which may be either ventral or lateral. A few tadpoles have paired spiracles. Shortly after the operculum forms, the original gills degenerate and new gills develop from the walls of the gill clefts. Although they are enclosed in an atrial chamber, these are external gills derived from the integument. An opercular fold also forms in the salamanders and caecilians but it is small, consisting simply of a crease anterior to the gill region, so that no atrial chamber is formed.

The gills of anurans and caecilians are resorbed during metamorphosis, the gill slits close, and the lungs take over. Salamanders show more variety. They seem to be experimenting with different methods of solving the problem of "breathing" on land. Most terrestrial salamanders lose gills and gill slits (see Table 2-1) and acquire lungs just as the frogs do. Members of the family Plethodontidae never develop lungs. The vast majority are terrestrial and depend entirely on respiration through the skin and lining of the mouth and pharynx. The aquatic *Amphiuma* and *Cryptobranchus* develop lungs and lose their gills, but retain the openings of one pair of gill slits. Adult Proteidae possess both gills and lungs and two gill slits remain open.

Adult Trachystomata have lungs, gills, and gill slits. If the gills of a Dwarf Siren (*Pseudobranchus*) are removed, it can survive by coming to the surface frequently to gulp air. If a Dwarf Siren is put in an aquarium with a screen to prevent it from coming to the surface, it expands its gills and continues to survive. The importance of the role of cutaneous respiration in sirens and the aquatic salamanders has not been determined. The Proteidae and Trachystomata are sometimes grouped as perennibranchs because both have gills as adults, but they are not closely related.

Lungs. The lungs of amphibians usually develop at the time of metamorphosis, sometimes earlier, and are relatively simple in structure. The left lung of caecilians is very short, and alveoli (sing. alveolus), little pockets at the ends of the tubes, are present only in the right lung. Salamander lungs are paired elongated sacs, the left one usually the longer. In some

TABLE 2-1.   DISTRIBUTION OF GILLS, GILL SLITS, AND LUNGS
AMONG ADULT CAUDATES AND TRACHYSTOMES

Family	Number of pairs of gills	Number of pairs of slits	Lungs
Hynobiidae	0	0	yes (except *Onychodactylus*)
Cryptobranchidae			
Cryptobranchus	0	1	yes
Megalobatrachus	0	0	yes
Ambystomatidae[1]	0	0	yes
Salamandridae	0	0	yes
Amphiumidae	0	1	yes
Plethodontidae[1]	0	0	no
Proteidae	3	2	yes
Sirenidae			
Siren	3	3	yes
Pseudobranchus	3	1	yes.

[1] Certain members of these families fail to metamorphose and retain the larval gills and gill slits when sexually mature.

salamanders, the lining is smooth, in others alveoli are present in the basal part. Lung linings are more complex in the frogs, which is what we should expect since they have been more successful, on the whole, in making the transition from gills to lungs than have the salamanders. The walls of their lungs are made up of many folds lined with alveoli. In general, the more terrestrial the frog or toad, the larger are the alveolar respiratory surfaces in the lungs.

Respiratory Passages. Nasal passages of the amphibians lead from the external nares (nostrils) to the internal nares or choanae, openings in the roof of the mouth just inside the upper jaw. Terrestrial amphibians have mechanisms for controlling the size of the external aperture, whereas many larval salamanders, the adult Sirenidae, and frog tadpoles have valves around the internal nares to control the direction of water flow. The mouth leads into the pharynx, a gateway chamber which passes into the esophagus of the digestive system and is connected to the lungs by a tube called the trachea. The trachea is short in most salamanders but in *Amphiuma* and the trachystome, *Siren*, it may be four or five centimeters long. Frogs generally have such a short trachea that it can hardly be said to exist, although one is definitely present in the aquatic Pipidae in which the lungs act as hydrostatic organs. The trachea divides at the lower end into two bronchial tubes which lead to the lungs.

During aquatic respiration water passes from the mouth to the pharynx, and out the gill slits. In salamanders and trachystomes this makes a current of water which flows over the gills. Water enters the atrial chamber of anuran tadpoles, flows over the gills, and passes to the outside through the spiracle. Amphibians breathing air keep the mouth tightly closed and draw air in or push it out through the nasal passages by lowering or raising the floor of the mouth.

Larynx. Since amphibians are the lowest form of vertebrates that can hear it is not strange that they are the first ones to have the anterior end of the trachea modified to form a voice box or larynx. The larynx opens into the pharynx by a slit-like glottis, bounded on each side by a bar of cartilage derived from the last visceral arch. These cartilages are usually divided into upper (aretynoid) and lower (cricoid) ones. Sometimes the cricoid cartilages fuse to form a ring. Frogs and toads have two muscular bands, the vocal cords, stretching across the larynx parallel to the glottis. Air passing over the vocal cords makes them vibrate to produce sounds. Tightening or relaxing the vocal cords results in variations in pitch.

Voice is well developed in the anurans. A female frog may grunt when held in the hand or give a "mercy scream" when caught by a snake. The calling of a chorus of breeding males is one of the most familiar of all animal sounds. Some frog species may lack voices, but not all of those reported to be silent are indeed so. *Eleutherodactylus cundalli* of Jamaica was for many years considered a voiceless, woodland species. Recently it has been found to call in limestone caves and crannies in rocks rather than at the surface.

Salamanders lack vocal cords and most of them are voiceless. Plethodontids also lack trachea and larynx. A few salamanders have "voices" but the sounds they produce seldom amount to more than faint squeaks or grunts. The Pacific Giant Salamander, *Dicamptodon ensatus,* has a low-pitched, rattling note, and makes a screaming sound when in danger. The trachystome genus *Siren* was so named because it was originally reported to "climb trees and quack like young ducklings." It does neither, but the name still persists.

Vocal Sacs. The males of many frogs have vocal sacs, out-pouchings from the mouth cavity extending ventrally and laterally under the skin and muscles of the throat. Sometimes there is a single vocal sac, sometimes a pair of them. They may be very large and extend posteriorly into the large lymphatic spaces. When a frog calls, its vocal pouch is filled with air and may swell out as a glistening bubble which collapses at the end of the call. How effective these pouches are as resonators is clear to anyone who has héard the bellowing of a Bullfrog (*Rana catesbeiana*) or the ear-splitting blast of a Great Plains toad (*Bufo cognatus*). The calls of different species of frogs are just as characteristic as the songs of birds. Furthermore, individual frogs are apparently able to vary the pitch of the call. Some species of frogs have a "rain call" which differs from the mating call given at the breeding ponds.

Circulatory System

In conjunction with the development of lungs, the amphibians also evolved a double circulatory system comprising a pulmonary circulation and a systemic circulation.

Heart. Instead of the simple two-chambered heart characteristic of most fishes, amphibians have a three-chambered heart with a right atrium, a left atrium, and a single ventricle. Blood coming back from systemic circulation empties into the right atrium. Pulmonary veins empty into the left atrium. The lungless plethodontids lack pulmonary veins and their left auricle is small. The amphibian ventricle is not divided but nevertheless the two bloodstreams, pulmonary and systemic, mix only to a slight degree. Some amphibians have a rather complicated system of valves and partitions that helps keep the two streams separate.

Arteries. In the primitive ancestor of the vertebrates, blood apparently left the heart through a large artery running forward, the ventral aorta, which gave off six pairs of branches, the aortic arches. These arches curved around the gut to unite on the dorsal side and form the dorsal aorta, the main systemic artery distributing blood to the body. The history of the evolution of the arterial system of the vertebrates has been largely a history of the reduction of the aortic arches. Most fishes have lost the first two arches. The last four are divided into afferent and efferent parts joined by the capillaries passing through the gills. In larval amphibians and the gilled salamanders the aortic arches are not interrupted by gill capillaries. Instead the arches give off vascular loops which pass into the gills and branch into capillaries, which are gathered together again to rejoin the aortic arches dorsally. This simplifies the change-over from gill to lung respiration at metamorphosis. During this stage the vascular loops degenerate, the arches expand, and the flow of blood from heart to body continues without interruption.

Salamanders are more primitive than frogs in the evolutionary stage reached by their aortic arches. The first two arches are lacking; the forward extensions of the ventral aorta from which they originally arose form the

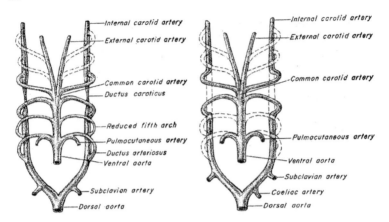

FIG. 2-3. The condition of the aortic arches as found in most salamanders (left) and frogs (right), ventral view.

external carotid arteries carrying blood to the jaws. The third pair of arches run forward as the internal carotid arteries to supply the face and brain. Frequently a connection remains between the third and fourth arches. The fourth pair are the systemic arches which form the main part of the dorsal aorta. The fifth arches may persist although, if so, they are much reduced in size. The sixth arches give rise to the pulmocutaneous arteries which carry blood coming mainly from the right side of the heart to the lungs and skin to be oxygenated; each retains its connection to the arms forming the dorsal aorta through a ductus arteriosus. Thus some of the unoxygenated blood passing through the sixth arches may go to the body rather than to the lungs—not a very efficient arrangement. In frogs there is never any connection between the third and fourth arches, the fifth arches have disappeared, and the sixth pair have lost their connections with the dorsal aorta so that the unoxygenated blood they carry must go to the lungs or skin.

Veins. Like the other vertebrates, amphibians have an hepatic portal system. The veins carrying blood from the intestine join to form the hepatic portal vein which branches into a network of small sinuses in the liver. These vessels are then gathered together again into the hepatic veins to return blood toward the heart. The liver performs important chemical functions in the body. It removes food materials that have been absorbed by the blood from the intestine, changes them chemically, stores some in the form of glycogen, returns others to the blood in a form that can be used by the cells. It also removes the harmful nitrogenous waste products of cell metabolism, changes them into harmless urea, and returns this to the blood to be removed by the kidneys. Obviously, it is to the advantage of an animal to have as much blood as possible pass through the liver to be cleansed and restocked with food before its return to the heart. In this the amphibians show an advance over the fishes. The anterior abdominal vein carrying part of the blood returning from the hind legs and posterior part of the body discharges directly into the hepatic portal vein. The rest of the blood from the hind legs passes through the renal portal system of the kidneys. Amphibians also have pulmonary veins, which are lacking in fishes except for lungfishes.

Excretory System

An animal must have some way of getting rid of the waste products of metabolism and of regulating the salt content and water balance of the body. These are the functions of the excretory system, which consists of the kidneys and the tubes leading from them to the outside. The excretory and reproductive systems are closely connected, especially in males, so that it is difficult to discuss one without referring to the other. This connection is apparently an accident of development with but debatable functional advantage.

Kidney. Three successive types of kidneys are characteristic of stages in the evolution of the vertebrates. The kidney of the adult amphibian, as in most fishes, is a mesonephros,[1] midway in the evolutionary series between the primitive, anterior pronephros of the hagfish and the compact, posterior metanephros of the higher tetrapods. The kidneys of caecilians are long and extend the length of the body cavity. They are also long in salamanders and are divided into two parts: an expanded posterior part, and a narrow anterior part the tubules of which, in males, carry sperm from the testis to the mesonephric or Wolffian duct (ductus deferens). This duct, into which the collecting tubules of the kidney also drain, runs along the outside of the kidney to empty into the cloaca. The kidneys of frogs occupy a more posterior position and are flattened dorsoventrally. They are not divided into anterior and posterior parts and they enclose the Wolffian ducts. As in salamanders, tubules from the testes pass through the kidneys.

Bladder. Amphibians have a thin-walled urinary bladder which opens into the cloaca and is not connected with the Wolffian ducts from the kidneys. Urine passes down these ducts directly into the cloaca and then backs

[1] In some recent texts the term mesonephros is restricted to the kidney structure of the embryonic amniotes and the term opisthonephros is used for the adult kidney structure of lampreys, most fishes, and the amphibians.

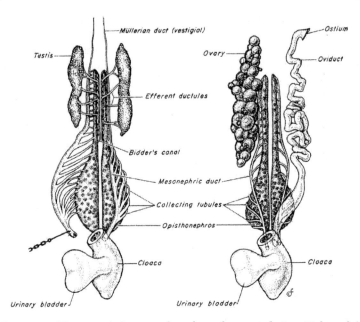

FIG. 2-4. The urogenital organs of a salamander, ventral view. Male on left, female on right.

up into the bladder for storage. The liquid discharged by a captured toad on the hands of the collector. is not urine, but reserve water stored in the cloaca. It has been suggested that the action lightens the weight of the body and so facilitates the escape of the animal.

Reproductive System

In all living things the two critical concerns are the maintenance of the individual and the perpetuation of the race. The second is the function of the reproductive system. As is typical in vertebrates, the amphibian reproductive system is made up of primary sex organs, the gonads, and accessory organs, including ducts and other structures concerned with bringing the germ cells together.

Female Reproductive Organs. The amphibian ovary is saccular and has an enclosed lymphatic cavity. In vertebrates, the size of an individual ovum is largely determined by the amount of contained yolk, the material from which the developing embryo draws its nourishment. Amphibian eggs have a moderate amount of yolk: more than the microscopic eggs of most mammals in which the embryo receives nourishment from the maternal bloodstream, but less than the large-yolked eggs of birds in which the young reach an advanced state before hatching. The eggs of many amphibians are about two millimeters in diameter and may be very numerous. Frogs in particular may produce thousands of mature ova at one time. Ovaries filled with ripe eggs are irregular lumpy structures filling the greater part of the body cavity. The eggs escape into the coelom through the external wall of the ovary.

Ovaries vary in shape with variation in body form. Those of caecilians, are long and narrow. Salamander ovaries are also elongated, but not as much as those of caecilians. Frogs, on the other hand, have decidedly foreshortened and more compact ovaries. The cavity of the salamander ovary is single and continuous, but in the frog ovary it is divided into a number of pockets.

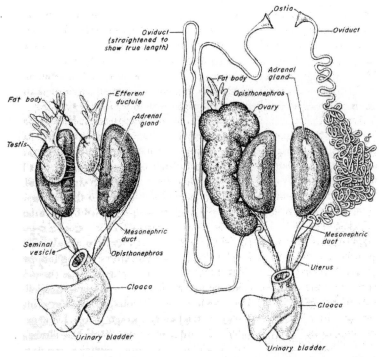

FIG. 2-5. The urogenital organs of a frog, ventral view. Male on left, female on right.

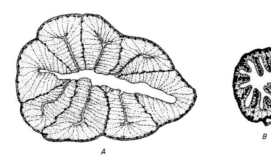

FIG. 2-6. Cross section of oviduct of a salamander, *Eurycea bislineata* (A) during the breeding season (April 11), and (B) after the breeding season (June 21). [From Weichert, *Anatomy of the Chordates*, McGraw-Hill, 1951, used by permission.]

Associated with the ovaries of amphibians are fat bodies or corpora adiposa. The fat bodies of salamanders are long, slender structures running parallel to the ovaries along their median edges. Those of frogs lie just anterior to the gonads and consist of several yellowish, fingerlike processes. That they apparently serve as storage places for nutritive material for the developing ova is indicated by their changes in size during the year. They shrink as the eggs enlarge, being smallest at the approach of and during the breeding season, after which they gradually increase in size until the maximum is reached just before the rapid growth of the ova begins.

The tubes that transport the products of the ovaries to the outside are the paired Müllerian ducts. These have the same general pattern for all amphibians. Anteriorly, each tube begins as an expanded, funnellike opening, the ostium, situated well forward in the body cavity. Eggs that have escaped from the ovaries into the body cavity are directed toward the ostium by cilia located on the peritoneum of the body wall, on the liver, and on adjacent structures. Over most of its length the Müllerian tube has a thickened wall with a glandular lining. During the breeding season, the duct becomes greatly enlarged and markedly coiled and the glandular lining epithelium secretes a clear, gelatinous substance. After entering the ostium, the egg is forced along by peristaltic waves. As it passes down the oviduct with a twisting, spiral motion, the glands deposit several layers of jellylike material around it. Posteriorly, each oviduct is enlarged to form a uterus-like structure. These serve only as temporary storage places for eggs that are soon to be laid. The uteri of most amphibians enter the cloaca separately, but in *Bufo*, the two uteri unite and have a common opening into the cloaca. The female salamander possesses a dorsal diverticulum of the cloaca, the spermatheca, which serves to store sperm until the eggs are ready to be fertilized.

Male Reproductive Organs. The primary male reproductive organ is the testis, a compact structure made up of a mass of seminiferous (sperm-bearing) tubules which connect, by means of ducts, to the outside. Each testis is really a compound tubular gland. Its shape, like that of the ovary, is roughly correlated with body shape. The testes of caecilians are elongated structures that look like strings of beads. The beadlike swellings consist of masses of seminiferous tubules. These are connected by a longitudinal collecting duct. Salamander testes are shorter and more irregular in outline. The testes of frogs are compact and oval or nearly rounded in shape. There is a pronounced difference in the size of the testes during the breeding and the nonbreeding seasons. Like the ovaries, the male gonads also have fat bodies associated with them. They resemble the fat bodies of the ovaries in position and in appearance and, like them, fluctuate in size with the onset and passing of the breeding season.

Male toads of the family Bufonidae have a peculiar structure, Bidder's organ, that lies anterior to the testis, between it and the fat body. If the testes are removed, these lobelike structures will develop into functional ovaries in about two years, thus bringing about a complete reversal of sex in the male. When this happens, the otherwise rudimentary oviducts enlarge, seemingly in response to female sex hormones elaborated by the transformed Bidder's organs. This is a striking example of the fact that each sex of the vertebrates has rudimentary organs of the other.

There is an extremely close relationship between the reproductive and excretory systems of the male amphibian. The Wolffian duct carries the products of both kidney and testis to the outside and the tubules through which the sperm are transported to the Wolffian duct pass through the anterior part of the kidney.

The longitudinal collecting duct of the caecilian testis gives off small transverse canals between the successive lobules. These, in turn, pass toward the kidney to join another longitudinal duct running along the median edge of the kidney. Spermatozoa pass through this duct into a second series of transverse canals that join the mesonephric tubules. The tubules then enter the Wolffian duct.

In the salamander, the efferent ductules carrying sperm from the testis join a fine longitudinal canal, called Bidder's canal, that runs along the median edge of the kidney. Bidder's canal connects by means of a number of short ducts with the mesonephric tubules in the anterior portion of the kidney. (Bidder's canal is present, though rudimentary, in the female as well.) The tubules emerge from the lateral edge of the kidney to join the Wolffian duct. Thus the anterior end of the Wolffian duct is concerned primarily with the passage of spermatozoa, whereas the posterior portion serves the dual purpose of elimination of urinary waste and transportation of spermatozoa. The Wolffian ducts of the two sides enter the cloaca independently.

The urogenital ducts of frogs are quite similar to those of salamanders, but there are minor variations. The efferent ductules enter the anterior end of the kidney along its median edge. In some forms, they connect directly with the Wolffian duct, but in others they join a Bidder's canal lying close to the median border but within the kidney. Spermatozoa are then transferred from Bidder's canal through the mesonephric tubules to the Wolffian duct. A Bidder's canal is present in the female, but its function, if any, is unknown. The Wolffian duct emerges from a point near the posterior end of the kidney and passes into the cloaca. Males of many species of frogs have a dilation in the Wolffian duct close to the junction with the cloaca. This forms a seminal vesicle for the temporary storage of sperm. Such seminal vesicles are poorly developed in *Rana pipiens* and *Rana catesbeiana*, the two species most commonly studied in introductory zoology courses. Seminal vesicles, when present, are largest during the breeding season and shrink after breeding is over. Such seasonal changes indicate that their development and function are under hormonal control.

The Wolffian duct, particularly that of salamanders, also shows seasonal variation in size, becoming large at the height of reproductive activity. In the nonbreeding season, it may be reduced to a mere thread, particularly anteriorly where it functions solely as a reproductive duct.

Males of all known salamanders except the Hynobiidae and Cryptobranchidae undergo an enlargement of the glandular lining of the cloaca as the breeding season approaches. The glands secrete a jellylike material around a cluster of sperm to form a packet which rests on a jelly base also secreted by glands of the cloaca. Together they form a toadstool-shaped structure, the spermatophore.

Skeleton

The skeleton of a vertebrate is that part of the body made up of cartilage and bone. Bone may develop either in the skin (the dermal skeleton or exoskeleton) or deep within the body (the endoskeleton). The endoskeleton is laid down first as cartilage which is replaced more or less completely during development by bone. Dermal bone forms directly with no cartilaginous precursor. Generally, when we speak of the skeleton of an animal, we mean the endoskeleton, along with some parts of the exoskeleton, particularly in the skull region, which have dissociated themselves from the skin and sunk down to join the underlying endoskeleton.

The main function of the exoskeleton is to protect the body. The endoskeleton provides a place of attachment for the muscles, a framework against which they can pull in moving the body. When the vertebrates moved out on land, a third function of the skeleton became very important. The bodies of aquatic animals are supported by the water in which they live. Animals living in the much thinner medium of air lack this support. Their limbs are no longer mere steering fins; they must bear the weight of the body as it moves across the ground or through the air. Their backbones must support their internal organs. At the same time, success in the more complex environment on land demands a diversity of movement unnecessary for the aquatic ancestors of the tetrapods. The skeletons of land vertebrates are more completely ossified, more solidly put together, and the articulations are more complex than are those of fish. We see the beginning of these changes in the amphibians.

Except for some of the bones of the skull, the functional exoskeleton has almost disappeared in the amphibians. Certain vestiges (scales, claws) were discussed in the section on the integument.

Skull. The skull of a modern amphibian is flattened and has fewer bones than does a fish skull. The primitive cartilaginous skull box, the chondrocranium, has been partly replaced by bone. The attachment of the jaw is autostylic, that is, the upper jaw is connected directly to the skull. (The hyostylic jaw found in many fishes is braced against the skull by the hyomandibular bone, derived from the second visceral arch.)

Vertebral Column. The structure of the individual vertebra will be discussed in Chapter 4. The vertebral column of fish consists of two sections: in front of the region of the anus are trunk vertebrae; behind it are caudal vertebrae. Each of the caudal vertebrae has a haemal arch, a V-shaped structure below the centrum through which the caudal blood vessels pass. The vertebral column of amphibians shows further differentiation. The first vertebra is modified into a cervical or neck vertebra having two concave facets for articulation with the skull. The pelvic girdle articulates with the transverse processes of the last of the trunk vertebrae, known as the sacral vertebra. This still does not provide for very strong support by the hind limbs and most amphibians do not lift the body off the ground and truly walk on their legs. Salamanders wriggle and frogs hop. Loss of the tail in adult frogs has, of course, involved a loss of caudal vertebrae. Their place is taken by a long, rodlike structure, the urostyle.

Ribs. Two-headed ribs are present, though poorly developed, in modern amphibians. They are longest in caecilians, shorter in salamanders, and in most frogs are lacking entirely or are only present as cartilaginous tips on the transverse processes of the vertebrae.

Sternum. The sternum, or breastbone, which is characteristic of the higher tetrapods, appears for the first time in the amphibians, although not in a well-developed state. The ribs never attach to it to form a definite thoracic basket. It is absent in caecilians and some of the elongated salamanders. Other salamanders have a small triangular plate lying between the halves of the pectoral girdle. The frogs, as usual, are more advanced than the salamanders and have a definite sternum.

Girdles and Limbs. The appendicular skeleton of the tetrapods comprises a pectoral girdle to which the forelimbs are attached and a pelvic girdle to which the hind limbs are attached. The pectoral girdle does not articulate with the backbone. That of the amphibians, particularly the salamanders, is largely cartilaginous rather than ossified. The pelvic girdle articulates with the sacral vertebra. It is well developed in the anurans to provide attachment for some of the powerful muscles of the hind limb.

The limbs of amphibians, as of all tetrapods, are modifications of a basic pentadactyl (five-digit) plan. For the salamanders these modifications consist mainly of fusion of some of the carpal bones and loss of one or more of the digits. Frogs have radius and ulna, and tibia and fibula fused. The bones of the tarsus are elongated thus producing a hind leg specialized for jumping. A small additional bone, the prehallux, frequently occurs on the inner side of the foot.

Sirens are without hind limbs and caecilians lack both limbs and limb girdles.

Endocrine Glands

There is little to be said here about the structure of the endocrine glands. Most of them are found in all vertebrates and the morphological variations from one group to another seem in general to have little evolutionary significance.

Parathyroids appear for the first time as definitive structures in amphibians and seem to be concerned with controlling the level of calcium salts in the blood. They are present in all higher vertebrates and their removal results in death.

The two component parts of the adrenal gland are closely associated in

FIG. 2-7. Photograph of salamanders, *Ambystoma texanum*. (Lower) Normal control animal; (center) a few hours after removal of pituitary gland, pigment cells contracted; (upper) several weeks after pituitary removal; the dark color is due to failure of the animal to shed the corneal layer of epidermis, which has become very thick. The last effect is undoubtedly caused by the failure of the thyroid gland to function in the absence of the pituitary gland. [From Weichert, *Anatomy of the Chordates*, McGraw-Hill, 1951, used by permission.]

amphibians as in reptiles and birds, instead of being separated as in the lower vertebrates. The reason for this difference is unknown.

Although morphologically the endocrines offer little of interest, functionally they are among the most fascinating of organs. The thyroid of amphibians is the gland controlling metamorphosis. If it is removed from a tadpole, the tadpole grows into a large, fat, super tadpole, but never changes into a frog, although it does develop lungs and reproductive organs. If thyroid extract is fed to such a tadpole, it metamorphoses. On the other hand, if a tiny tadpole is fed thyroid extract it stops growing and quickly metamorphoses into a midget frog. The thyroid gland also controls ecdysis. An amphibian from which this gland has been removed does not shed its skin and the dead layers pile up until the animal appears much darker than normal. When fed thyroid extract it soon casts off the dead epidermal layers. One function of the pituitary gland is apparently to control the thyroid because

removal of the pituitary results first in a loss of color (due to contraction
of pigment cells) and then in a darkening caused by failure to shed the skin,

Nervous System

Animals on land face a far more complicated environment than do those
in water. To cope with it, the higher vertebrates have developed a more
complicated nervous system. The chief evolutionary change in the vertebrate brain is the enlargement of the cerebral hemispheres, lateral lobes of
the forepart of the brain. Primitively, the roof and sides of these hemispheres are composed of a nonnervous epithelial layer, the pallium. The
amphibians have made few important advances over the fish in the structure of the nervous system, but in them for the first time scattered nerve
cells appear in the wall of the pallium, foreshadowing the enormous development of the cerebrum in mammals.

Like fishes, amphibians have only ten cranial nerves, not twelve as do
the higher vertebrates. The movements of tetrapod limbs are more complicated than are those of fish fins and require a more complex innervation.
Enlargements of the spinal cord in the cervical and lumbar regions result;
these appear first in the amphibians. The spinal cord of the salamanders
extends to the tip of the tail but that of frogs is much shortened and the
spinal nerves continue through the neural canal as a brushlike structure,
the "horse's tail" or cauda equina.

Sense Organs

Eye. The cornea, the clear window in the outer layer of the vertebrate
eyeball through which light rays enter, becomes opaque when dried. Animals that live on land must have some mechanism to keep the cornea moist
and to wash it free of specks of dirt. Terrestrial amphibians have developed
a series of glands in the tissue around the eye and movable eyelids to wash
the secretions of these glands across the eyeball. The eyeball can be protruded or withdrawn into the eye socket. When it is drawn in, the lower
lid moves up to close over it. Permanently aquatic salamanders and sirens
lack eyelids as do all amphibian larvae. Eyes are very poorly developed
in the caecilians and in blind, cave dwelling salamanders.

Ear. Fishes perceive low-frequency vibrations by means of a series of
sense organs arranged in rows on the head and sides of the body, the lateral
line sense organs. Lateral line systems still persist in larval amphibians,
aquatic salamanders, and the trachystomes, but these sense organs can func-

tion only in water. When the amphibians moved out on land they evolved a wholly new sense organ to receive low-frequency vibrations in the air—the act that we call hearing. This new sense organ developed as a part of the inner ear which heretofore had only transmitted to the brain information about the position and movements of the body. Two small outpocketings from the inner ear of the amphibians, the lagena and pars basilaris, are believed to be concerned with hearing. The lagena is the forerunner of the marvellously complex cochlear duct of the higher vertebrates. Frogs develop a middle ear from the first pharyngeal pouch. A tympanic membrane, lying flush with the surface of the head, picks up vibrations from the air and transmits them to the columella, a small rod-shaped structure lying in the tympanic chamber (cavum tympanum) behind the membrane. The columella passes the vibrations on to the inner ear where the nerve receptors are located. A Eustachian tube connects the tympanic chamber with the pharynx and serves to equalize air pressure around the tympanic membrane. Some burrowing toads and all salamanders lack middle ears. Aquatic salamanders seem to receive vibrations to the inner ear by way of the jaw and the bones which connect it to the cranium; terrestrial salamanders, by way of the forelimb and shoulder girdle. The "hearing" done by such means must be very limited. Perhaps this is why the salamanders have lagged so far behind the frogs in developing voices.

Jacobson's Organ. This sense organ appears for the first time in the amphibians. It is also well developed in lizards and snakes but is vestigial in most of the other tetrapods. In amphibians it consists of a pair of blind sacs connected by ducts to the nasal cavities. Since it is inervated in part by a branch of the olfactory nerve it probably plays a part in the recognition of food.

Collateral Reading and General Reference

Boulenger, G. A. *The Tailless Batrachians of Europe.* Pt. I. London: Printed for the Ray Society, 1897.

Ecker, A., R. Wiedersheim, and E. Gaupp. *Anatomie des Frosches.* 2d ed., Braunschweid, 3 vols., 1888–1904. (Old but still the definitive work on frog anatomy.)

Francis, E. T. B. *The Anatomy of the Salamander.* London and New York: Oxford University Press, 1934. (An excellent, detailed study of the anatomy of a single species of salamander.)

Holmes, S. J. *The Biology of the Frog.* 4th ed. New York: McMillan, 1927. (A standard work on frog anatomy.)

Noble, G. K. *The Biology of the Amphibia.* New York: McGraw-Hill, 1931. Reprinted by Dover Publications, 1954. (Somewhat out-of-date but still by far the best reference work on the biology of the amphibians.)

Weichert, C. K. *Anatomy of the Chordates.* 2d ed. New York: McGraw-Hill, 1958. (All standard comparative anatomy texts include much information on the structure of amphibians.)

Young, J. Z. *The Life of Vertebrates.* London: Oxford University Press, 1950. (A comprehensive, modern treatment of many aspects of the biology of the vertebrates.)

STRUCTURE OF
REPTILIA

TWO GREAT ADVANCES make the reptiles more successful invaders of the land than are their more primitive amphibian relatives. The first of these is the amniote egg. As the embryo develops, folds of tissue grow up around it and fuse above it to form two closed sacs. The outer one, the chorion, surrounds the whole egg with a protective membrane. The inner one, the amnion, is filled with fluid to provide the embryo with an individual aquatic environment in which it can develop without danger of desiccation. A third extraembryonic membrane is the allantois, a saclike structure that grows out from the hindgut of the developing embryo to press against the chorion. It acts as a respiratory organ and also as a place for the storage of the waste products of metabolism. A fourth membrane develops around the yolk to form the yolk sac. The entire complex of developing embryo and extraembryonic membranes is surrounded by a watery albumen (except in snakes and lizards) and is covered with a tough shell. Since the reptilian eggshell is permeable, some environmental moisture is necessary to prevent desiccation. But the egg is much better adapted to deposition on land than is the unprotected egg of even the most terrestrial amphibian.

The second advance that makes the reptile the success it is on land is the presence of epidermal scales. The amphibian epidermis is merely a soft secreting structure which offers little protection against desiccation or other hazards of life in the open air. The epidermis of reptiles practically ceases to function as a secreting membrane and becomes a protective structure, guarding the animal from actual physical injury as well as acting as a mem-

37

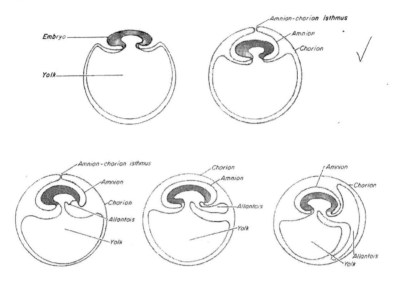

FIG. 3-1. Diagram showing the formation of the amnion and chorion in the developing amniote egg.

brane to shield it from desiccation. Epidermal scales are also found on the legs of birds and on some mammals (e.g., armadillos).

DEFINITION OF REPTILIA

The reptiles—turtles, lizards, snakes, crocodilians, and tuatara—are amniotes that have epidermal scales but lack feathers, hair, or mammary glands. Development is direct since the need for a larval stage ended with the evolution of a shelled egg that could be laid on land. The stratum corneum is much better developed than that of amphibians and the skin is dry. In addition to heavy epidermal scales, many reptiles have bony dermal plates lying under the epidermis. There are almost no skin glands. Respiration, as would be expected, is entirely by means of lungs, except for some that takes place through vascular tissue in the cloaca and pharynx of aquatic turtles. A secondary palate, which in higher amniotes separates the nasal cavity from the oral cavity, begins to develop in the reptiles so that at least a partial separation occurs. In crocodilians this palate becomes complete. The regional differentiation of the vertebral column characteristic of mammals makes its appearance in the reptiles, though only the crocodilians have five clear-cut regions. There is a single occipital condyle.

The legs usually have five digits ending in true claws. The kidney is meta-nephric like that of birds and mammals. The heart is either three-chambered or, in the crocodilians, four-chambered. There are twelve pairs of cranial nerves. Like the amphibians, reptiles are ectothermic.

ANATOMY OF REPTILES

Present-day reptiles do not match the birds or fishes in number of species or of individuals. They have, however, successfully invaded many habitats so that they now occupy the seas, fresh water, land, trees, and soil; the flying dragons (*Draco*), have even become gliders. For so small a group the reptiles show a wonderful variety of structural modifications.

Because reptiles are so diversified it is difficult to give a concise account of their structure applicable to all. The following is a "ground plan" to serve as a basis for understanding the modifications that characterize the different groups discussed in later chapters.

Integument

Since the skin of a reptile does not serve as a respiratory organ, it does not need to be kept moist. The important problem now is to cut down on water loss. That the dry, tough, scale-covered skin of the reptiles performs this function well is shown by the number of snakes, lizards, and even turtles that have adapted themselves to desert conditions and are able to go for long periods without water. (Some reptiles apparently need no other water than that obtained from their food though most do drink when water is available.)

Epidermis. The stratum corneum of the epidermis, which appeared first in the amphibians, is much better developed in the reptiles and serves admirably to protect them from desiccation. As in amphibians, this layer is sloughed off periodically in ecdysis.

Dermis. The dermis is well developed and in many snakes and lizards has an abundance of chromatophores. That the dermis of some reptiles makes an admirable leather is attested by the popularity of alligator pocketbooks and snakeskin shoes.

Glands. The many epidermal glands that keep the skins of amphibians moist have practically disappeared in the reptiles, and the amount of water lost from the body surface is therefore much less. Most of the few remaining skin glands produce strong-smelling substances for sexual attraction. Both male and female crocodilians have two pairs of musk-secreting glands, one

pair lying on the inner halves of the lower jaw and the other just inside the
cloacal opening. Crocodilians also have a row of glands down either side of
the back between the first and second rows of dermal plates. Their function,
if any, is unknown. The so-called femoral and preanal "glands" on the under-
surfaces of the thighs in male lizards are apparently not glands at all. They
are small pits filled with a yellowish debris which piles up in small, hard
cones possibly used to help clasp the female in copulation. Some snakes
have glands in the cloacal opening; the nauseating odor of their secretion is
probably protective. Some turtles have musk glands along the lower jaw
and in the line between the plastron and the carapace. Most likely they are
used for sexual attraction. But let no man who has succumbed to the allure
of *Nuit d'Amour* sneer at the lowly Stinkpot Turtle.

Scales. The characteristic scales of reptiles are not homologous with the
dermal scales of fishes but are different structures derived from the epidermis. However, some reptiles also have true dermal scales. They are
especially well developed in turtles and fuse with each other to form the
heavy plastron, or belly shell, and with the ribs and vertebrae to form the
carapace, or back shell. The crocodilian dermis is thick and soft except on
the dorsal side of the body and occasionally under the throat where small
bony dermal scales lie beneath the epidermal scales. Crocodilians and
Sphenodon also have gastralia, dermal bones lying in the ventral body wall
between the true ribs and the pelvis. Some lizards have dermal scales
(osteoderms) underlying the horny epidermal scales.

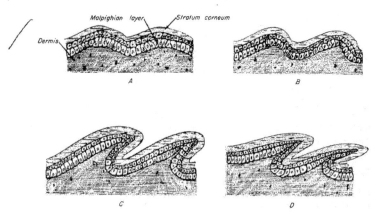

FIG. 3-2. Diagram showing the stages in the development of epidermal reptilian scales. [After Weichert.]

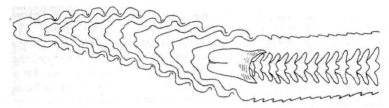

FIG. 3-3. The structure of the rattle of a rattlesnake. [After Garman.]

The epidermal scales of reptiles are of two sorts; one is characteristic of the snakes and lizards, the other of turtles and crocodilians.

Each scale of a snake or lizard projects backward to overlap the one behind. The scales are formed from thickened areas of the integument which grow up and backward and become cornified; they are continuous with each other at their bases. Periodically a new set of scales forms beneath the old and when this happens the outer, older series is shed. Snakes turn the skin inside out as they shed it. Lizards simply creep out of the old skin, leaving it right side out, or shed it in flakes (see Fig. 3-2).

The outer portions of the so-called horns of the horned lizard (*Phrynosoma*) are modified epidermal scales as are the rattles of the rattlesnake. At the time of ecdysis the scale at the extreme tip of the tail is not shed with the others but is held in place by a bump on the newly formed scale in front. Thus a new and larger section is added to the rattle each time the snake sheds its skin (see Fig. 3-3).

Crocodilians and turtles have epidermal scales, each of which develops separately so that the scales do not form a solid sheet. Crocodilian epidermal scales cover the entire body. Instead of being shed periodically, these scales wear away and are gradually replaced.

Most turtles have large epidermal scales that do not conform in pattern to the underlying bony dermal plates of the carapace and plastron, from which they are separated by a thin Malpighian layer (the basal layer of the epidermis). From time to time centers in the Malpighian layer cornify to form new scales, each slightly larger than the one above it. The old scales of some turtles peel off leaving the shell smooth, whereas on others they adhere so that the new ones are topped by little mounds of older, smaller scales and the shell is rough.

Digestive System

Glands. The oral glands, which help moisten the prey to prepare it for swallowing, are better developed in reptiles than in amphibians. A palatine gland is present on the roof of the mouth. Reptiles also have lingual (tongue), sublingual (below the tongue), and labial (lip) glands. The poison glands of the poisonous snakes are modifications of the labial glands in the upper jaw. There is one on each side which opens by a duct into the groove or cavity of the poison fang. The poisonous lizard *Heloderma* has a sublingual gland on each side that is modified to produce poison. Four ducts lead from each gland through the bone of the lower jaw to empty the poison into the vestibule in front of the grooved teeth. Thus while a rattlesnake can inject its poison at a single strike as with a hypodermic needle, *Heloderma* must hang on and chew to force the poisonous saliva into the wound. Marine turtles and crocodilians, which take their prey in water, have poorly developed oral glands.

Tongue. The tongue of turtles, crocodiles, and alligators is not protrusible and simply lies on the floor of the mouth. The Squamata have well-developed tongues (two-pronged in snakes and some lizards) which can be extended and retracted. Snakes have a small notch at the tip of the upper jaw so that even when the mouth is closed the tongue can be thrust out. Among lizards the chameleons have the best developed tongues, very useful in catching flies and other winged creatures.

Teeth. The history of the evolution of teeth has been one of reduction in number and localization of position. Fishes may have teeth almost anywhere in the mouth and in the pharynx; they may be on the tongue, the hyoid arch, and the gill arches, as well as on the jaws and roof of the mouth. Amphibian teeth are restricted to the jawbones, and to bones of the roof of the mouth—the palatines, vomers, and occasionally parasphenoids. Snakes and lizards have teeth on the palatines and pterygoids as well as on the jawbones, and *Sphenodon* has vomerine teeth. Crocodilian teeth are restricted to the jawbones as are those of the mammals and the extinct toothed birds. Turtles lack teeth but have instead sharp, strong, horny beaks.

Like the amphibians, the reptiles have teeth that may be replaced an indefinite number of times (polyphyodont) and that are usually the same all along the jaw (homodont). The hollow or grooved fangs of the poisonous snakes are modified teeth, the anterior teeth of some lizards are enlarged and canine-like, the lateral teeth of other lizards may be tricuspid or even oval crushing plates; these forms are thus heterodont rather than homodont. The teeth of snakes and lizards are either acrodont or pleurodont (attached to the side of the jawbone). Crocodilian teeth are set in sockets in the jawbone (thecodont), the first appearance of the typical mammalian condition.

Gut. The esophagus of reptiles is generally longer than that of amphibians and its walls are gathered into longitudinal folds to permit expansion during the swallowing of large prey. The stomach is clearly set off from the esophagus. Snake and lizard stomachs are long and spindle-shaped. Crocodilians have part of the stomach modified to form a muscular, gizzard-like region. The small intestine of reptiles is long and coiled. The large intestine usually has a colic caecum, a blind pouch at the juncture with the small intestine. Colic caeca are absent in crocodilians but are present in most birds and almost all mammals. The large intestine empties into a cloaca, which opens to the outside through the vent.

Acrodont   *Pleurodont*   *Thecodont*

Respiratory System

FIG. 3-4.
The types of tooth attachment found in amphibians and reptiles.

Since reptiles as a group are the first completely terrestrial vertebrates, they are the first to depend entirely on the lungs to aerate the blood. Cloacal and pharyngeal respiration supplement lung breathing in some aquatic turtles thus enabling them to stay under water for long periods. But in general reptiles are air breathers. Gills have disappeared completely and the lungs are better developed than are amphibian lungs.

Among the higher tetrapods, the anterior part of the respiratory tract is separated from the anterior part of the digestive tract. The separation is initiated in the reptiles in which a hard palate begins to form on the roof of the mouth. The openings of the internal nares are pushed to the back part of the mouth and the nasal passages between external and internal nares are elongated.

The reptilian larynx is little improved over that of the amphibians. Most reptiles are voiceless, but some lizards produce harsh sounds and alligators bellow vociferously during the mating season. A few turtles also have voices.

Circulatory System

When they became terrestrial, the reptiles were released from the necessity of circulating blood to the gills, but their consequent increased dependence on lungs called for improvements in the pulmonary circulation and heart.

Heart. The heart of most reptiles is three-chambered but crocodiles and alligators have achieved a completely four-chambered heart. The right atrium receiving unoxygenated blood from the body is completely separated from the left atrium receiving oxygenated blood from the lungs. Both atria of reptiles with three-chambered hearts empty into a single ventricle, but an incomplete septum growing from the apex toward the center of the ventricle essentially separates the two bloodstreams. The septum of crocodilians is complete.

Arteries. The third, fourth, and sixth aortic arches are present in reptiles, but their connections have been considerably modified. The truncus arteriosus, the large blood vessel that extends forward from the heart to give rise to the arches, splits at its base to form three main trunks. One of these, the

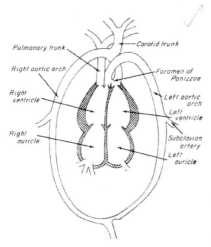

pulmonary aorta, leaves the right side of the ventricle and divides into two pulmonary arteries which represent the sixth aortic arches. The left (fourth) aortic arch has become connected with the right side of the ventricle and so carries unoxygenated blood. The rest of the truncus arteriosus connects with the left side of the ventricle and carries oxygenated blood. It crosses over the left aortic arch immediately after it leaves the heart and runs forward on the right side. First it gives off the right (fourth) aortic arch, then passes forward to divide into the common carotid arteries. These split into internal carotids, representing the third aortic

FIG. 3-5.

Diagram showing relation of aortic trunks to heart in the alligator.

arches, and external carotids, the forward extensions of the old ventral aorta. Thus the blood that goes to the head region is all oxygenated and the brain is assured of a good supply of oxygen. The right and left aortic arches fuse to form the dorsal aorta and since the former carries oxygenated and the latter unoxygenated blood, the blood going to the body is mixed even in the crocodilians with their four-chambered heart. In spite of improvements in the pulmonary circulation and heart, the reptiles cannot shunt all unoxygenated blood to the lungs and send only oxygenated blood to the body. That was left for the warm-blooded birds and mammals.

Veins. The venous system of the reptile shows little change from that of the amphibian. The large veins bringing blood back from the body are shifted over to empty into the right side of the auricle.

Excretory System

The reptiles are the lowest form of the vertebrates to have a true metanephros as the functioning kidney. The mesonephros appears only temporarily during development. An entirely new tube, the ureter, develops to carry the urine to the cloaca. During embryonic development it grows forward from the posterior end of the Wolffian duct to fuse with the collecting tubules of the metanephric kidney. The Wolffian duct now carries only reproductive products and gives rise to such masculine structures as the epididymus, ductus deferens, and seminal vesicles.

The bladder is an outgrowth from the cloaca. Snakes and crocodilians lack a bladder but most lizards and turtles have a well-developed, bilobed structure. As in amphibians, urine passes into the cloaca and then backs up into the bladder. Some turtles have accessory bladders which are used as organs of respiration or as reservoirs for water storage.

Reproductive System

The change that, above all others, made it possible for the reptiles to become truly terrestrial was the development of the shelled, amniote egg which could be laid away from water. Surprisingly enough, the structural modifications in the reproductive system that accompanied this revolutionary shift in habit were relatively minor. Perhaps most obvious are the copulatory organs of the males and even these apparently are not really necessary since *Sphenodon* and most birds achieve internal fertilization without the aid of such structures.

Female Reproductive System. Reptilian ovaries, like those of amphibians, are paired structures lying within the body cavity. The Squamata (lizards and snakes) have saccular ovaries as do the amphibians, but the other reptiles have solid ovaries as do the higher tetrapods. Reptile eggs are polylecithal, that is, they have a large amount of yolk, in contrast to the mesolecithal (moderate-yolked) amphibian eggs. This is because the young reptile hatches at a far more advanced state of development than does the larval amphibian and hence needs more nutriment to carry it along until it is able to get food for itself. Size of the ova depends largely on the size of the species, but all are considerably larger than amphibian ova. However, since reptiles have a much smaller number of ova (usually less than 100) that become mature at one time, the ripe ovaries, though sometimes very large, do not seem to dominate the body cavity as completely as they do in the frogs. Ovaries of *Sphenodon*, turtles, and crocodilians are rather broad and symmetrically placed, but those of the Squamata are elongated. This is particularly true of the snakes, in which the right ovary lies somewhat in advance of the left, an accommodation to the long, narrow body form.

Like those of the amphibians, reptile eggs break through the wall of the ovary to fall free into the coelom. The oviducts (Müllerian ducts) open into the coelom by narrow, slit-like ostia. Ova entering the ostium of the oviduct are forced along the tube by ciliary action and by muscular contractions of the wall. Each oviduct is differentiated into regions that perform different functions in depositing the envelopes surrounding the egg when it is laid. Fertilization must take place before these envelopes are formed. Since the cilia on the walls of the oviduct beat in a direction to move the egg from the ostium to the uterus, there must be some mechanism to allow the sperm to travel against the main current in order for them to reach the ovum soon after it enters the oviduct. Turtles have a narrow band of cilia along the side of the oviduct that beat toward, instead of away from, the ostium. This is presumably the path followed by the sperm to reach the ovum. Whether or not a similar path exists in other reptiles is not known.

Sphenodon, turtles, and crocodilians possess glands in the upper part of each oviduct for secreting albumen about the ovum. These glands are lacking in the Squamata and consequently their eggs lack albumen. The lower part of the oviduct, the so-called uterus, is specialized as a shell gland to secrete the shell around the egg. The two uteri enter the cloaca independently. The oviducts vary in size with the seasons, being largest at the height of breeding activity. The right oviduct of many reptiles, particularly the snakes, is longer than the left.

Since the reptiles have a metanephric kidney with a new duct, the old Wolffian duct of the mesonephros no longer has any function to perform in the females. It does persist as a vestigial structure, in close association with the ovary in snakes, turtles, and to a lesser extent in other reptiles.

Male Reproductive System. Like the amphibians, the reptiles have paired testes suspended within the body cavity. Usually they lie at about the same level but in snakes and lizards one is frequently farther forward than the other. The seminiferous tubules are long and coiled. The testes may be oval, round, or pyriform in shape; they fluctuate in size, as do those of amphibians, growing larger with the approach of the breeding season.

With the degeneration of the embryonic mesonephros the Wolffian duct no longer functions in excretion, but persists as a reproductive duct to transport sperm from the testis to the outside. The end of the Wolffian duct closest to. the testis is very much coiled to form part of a tubular mass called the epididymis. Persistent mesonephric tubules are modified into efferent ductules connecting the seminiferous tubules of the testis to the Wolffian duct. These ductules form the remainder of the epididymis which may be even larger than the testis. The Wolffian duct continues posteriorly as the ductus deferens, which is sometimes straight but more often convoluted. In most reptiles, the deferent duct on each side joins the metanephric ureter and with it enters the cloaca through a common opening at the tip of a urogenital papilla. The Müllerian duct commonly persists though generally it is reduced in size. The Müllerian ducts of the male European Wall Lizard, *Lacerta viridis,* are as well developed as are those of the female. Like the gonads, the epididymes and deferent ducts of reptiles so far studied show a seasonal modification in size and are apparently under endocrine control. Many lizards undergo a periodic enlargement of some of the posterior urinary tubules of the metanephros. These enlarged tubules produce an albuminous substance which, presumably, forms part of the seminal fluid in which the spermatozoa are suspended.

Snakes and lizards have glandular structures in the cloacal walls, whose secretion also contributes to the seminal fluid. No accessory glands are known in the turtles.

With the development of the amniote shelled egg, internal fertilization became mandatory. This is usually accomplished by copulatory organs. Two different types of copulatory organs are found in the reptiles, apparently representing separate evolutionary developments.

A single (occasionally multilobed), protrusible copulatory organ, the penis, is present in turtles. The penis is divided at its base, suggesting a double origin, but the distal end is single and free. There is a groove along the dorsal surface providing a passageway for the sperm. This penis is apparently derived from the thickened portions of the anterior and ventral walls of the cloaca and is made up of both connective and erectile tissue. The mass of erectile tissue is called the corpus cavernosum. During mating, the corpus cavernosum becomes filled with blood so that the penis grows firm and enlarged and can be extruded through the cloacal opening. This erection enables the penis to serve as an intromittent organ in the act of copulation.

The protrusible penis of crocodilians is longer and more slender than that of turtles and the groove is deeper. A spongy structure, the glans penis, at the outer end of the groove is homologous to the mammalian glans.

Snakes and lizards have paired copulatory organs, the hemipenes (sing.

hemipenis). The word means half a penis, but this is not really correct since each organ is separate and complete. The hemipenes are not homologous to the true penis of turtles, crocodilians, and mammals.

 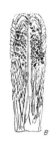

The hemipenes lie on either side of the base of the tail and form distinct thickenings so that it is frequently possible to determine the sex of a lizard or snake without dissection. Each hemipenis is a tubular structure which can be turned inside out like the finger of a glove. It bears a groove, the sulcus spermaticus, to transport the semen from the cloaca to the tip of the hemipenis. The distal end of the organ can be drawn back after copulation by means of a long retractor muscle. In a preserved specimen, the length of the hemipenis depends largely on the state of contraction of this muscle at the time of death. The external openings for the hemipenes can be seen on a snake by lifting the scale over the vent. The hemipenis of a lizard is usually short and broad and the inner surface (outer when the organ is everted) is typically pleated and folded. That of a snake

FIG. 3-6.
Representative hemipenes of snakes. A, Common Kingsnake, *Lampropeltis getulus*, southeastern United States; B, *Coluber hippocrepis*, Italy; C, Rosy Boa, *Lichanura roseofusca*, Lower California; D, Puff Adder, *Bitis arietans*, Africa.

is longer and the surface may be covered with spines and fingerlike projections arranged in rosettes called calyces (sing. calyx). These apparently serve to hold the hemipenis in the female cloaca during copulation. The hemipenis may be bilobed and the sulcus bifurcate.

In mating, usually only one hemipenis is inserted in the female cloaca. Which one is determined simply by which side the male happens to be on during copulation.

Skeleton

The skeletal modifications for terrestrial life originating in the amphibians are further developed in the reptiles.

Skull. The basic structure of the reptile skull does not differ greatly from that of the amphibian, although in general the bones are heavier and more completely ossified. Only one occipital condyle is present. Crocodilians have internal nares that open far back in the roof of the mouth and their nasal passages are separated from the mouth cavity by a bony secondary palate. The various modifications in the temporal region will be discussed in Chapter 5.

Vertebral Column. A typical reptilian vertebra consists of a ventral, spool-shaped centrum and, rising above it, a neural arch enclosing the spinal cord. Small, crescentic intercentra are often wedged ventrally between the centra of successive vertebrae. On each side of the anterior face of the neural arch is a process, the prezygapophysis, bearing an articular facet. This is a smooth surface which articulates with a similar facet on a similar process, the postzygapophysis, on the next vertebra forward. Each vertebra thus articulates with the next one at three points, the centrum and the two zygapophyses on the neural arch. Snakes and some lizards (not, strangely enough, the legless, snakelike ones) have two additional pairs of articular facets borne on structures at the base of the neural arch. The zygosphenes on the anterior face of a vertebra fit into grooves, the zygantra (sing. zygantrum) on the posterior face of the next vertebra forward. This makes a strong but pliable spinal column, a decided advantage in an animal that wriggles about as a snake does.

The first two vertebrae are modified to support the head and to allow for movement of the head on the neck. The first is called the atlas, since it bears the globe of the head as Atlas of mythology bore the world on his shoulders. The second, the axis, permits the head to be turned from side to side.

The vertebral column of mammals is divided into five regions characterized by different types of vertebrae. These include (1) cervical vertebrae

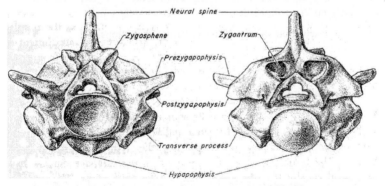

FIG. 3-7. Vertebra of a snake, *Eunectes murinus.* (Left) Anterior view; (right) posterior view.

lacking free ribs, (2) thoracics with ribs, most of which attach to the
sternum, (3) heavy lumbars without ribs, (4) fused sacrals attached to the
pelvic girdle, and (5) caudals. Reptiles are sometimes misleadingly said to .
show the same divisions. All of their vertebrae from cervicals to caudals may
have ribs. The vertebral column of snakes and legless lizards is most con-.
veniently divided into two regions. Anterior to the level of the vent is a
precaudal series of vertebrae bearing free ribs. Posterior to the vent is a
caudal series with the ribs fused to the vertebrae or absent. Usually
V-shaped chevron bones, probably remnants of the haemal arches of the
fishes, lie beneath the caudal vertebrae. Other lizards and *Sphenodon* have
cervical vertebrae whose ribs are usually short and do not attach to the
sternum; trunk vertebrae with longer ribs, the anterior ones attached to the
sternum; two sacral vertebrae with ribs attached to the pelvic girdle; and
caudal vertebrae, usually with chevron bones.

Turtles have cervical vertebrae that are rather loosely articulated to allow
the neck to be pulled back into the shell. The trunk vertebrae with their
ribs are fused to the carapace. There are usually two sacral vertebrae, joined
to each other and connected by ribs to the pelvic girdle. Like the neck
vertebrae, the tail vertebrae are not fused to the carapace.

Only the crocodilians have five easily distinguishable regions in the spinal
column. The cervical vertebrae have short ribs free from the sternum, the
thoracics have longer ribs with the anterior ones attached to the sternum,
there are two to five lumbars without free ribs, two sacrals attached to the
pelvic girdle, and a varying number of caudals.

The presence of two or more sacral vertebrae makes for better body support than is possible in the amphibians. Some reptiles are able to lift their
bodies off the ground and really walk—indeed many of the dinosaurs
walked solely on their hind legs.

Ribs. Along with the variations in the vertebral columns of the reptiles
goes a considerable variation in the number, structure, and attachment of
the ribs. They may be attached to the vertebrae by two heads or by a single
head, and they may articulate with the centra or to the neural arches. Furthermore, in a single animal the points of attachment may shift according to the
position of the vertebrae in the spinal column. At their ventral ends the ribs
of the anterior dorsal region usually join a sternum which is frequently
cartilaginous. Trunk ribs of *Sphenodon* and lizards may join a parasternum
lying between the sternum and the pelvic girdle. Turtles and snakes lack a
sternum. The trunk ribs of turtles are fused to the carapace. Snakes have
the ventral ends of the ribs connected with the belly scales. The posterior
cervical and anterior dorsal ribs of *Sphenodon* and the crocodilians each

bear a curved uncinate process which projects posteriorly to overlap the rib behind, giving strength to the thoracic body wall.

Girdles. Usually the pectoral girdle on each side is made up of coracoid, procoracoid, and scapula, with clavicle and interclavicle frequently present. The pelvic girdle is formed of three bones on each side, a dorsal ilium fused to the sacral ribs and a ventral pubis and ischium. The two pubic bones join ventrally in a pubic symphysis and the two ischia also form a symphysis. Between these two symphyses is a large, heart-shaped space, the cordiform foramen.

Nervous System

The spinal cord of reptiles, like that of the salamanders, extends the entire length of the vertebral column. Cervical and lumbar enlargements, made up of the cell bodies of the nerves that go to the limbs, are present except in the snakes and limbless lizards. The cervical enlargements of the turtles seem unduly massive. This is purely relative; the trunk muscles of these reptiles are so reduced that the portion of the cord between the enlargements, from which nerves pass to these muscles, is also reduced and is more slender than in other reptiles.

Some snakes and limbless lizards possess a faint, albeit fully developed, lumbosacral plexus, a network of nerves that ordinarily leads to the hind legs. This indicates that these animals must have arisen from ancestors that had legs.

Reptiles have twelve pairs of cranial nerves, as do all the higher tetrapods. The eleventh and twelfth pairs presumably represent the first two spinal nerves of the amphibians, indicating that a part of the spinal cord has been incorporated into the brain.

The cerebral hemispheres of the brain are larger than those of the amphibians, and a new area, the neopallium, appears in their roof. The migration of nerve cells to the outer wall of the neopallium, which takes place in the crocodilians, results in the formation of the first true cerebral cortex.

Sense Organs

On the whole, the major changes in the sense organs necessitated by terrestrial existence originated in the amphibians and have simply been further developed in the reptiles.

Eye. Snakes and some lizards lack movable eyelids but have instead a transparent window, the brille, in the skin covering the eye. It is this that gives snakes such a glassy-eyed, unwinking glare. They truly cannot close their eyes even though the eyes are always covered. Other reptiles have well-developed, movable lids, and a transparent, nictitating membrane, or third eyelid.

Ear. The auditory part of the inner ear is more highly developed in the reptiles than in the amphibians. Most reptiles have a tympanic membrane, usually lying flush with the head, and a middle ear cavity through which sound waves are transmitted to the inner ear. Snakes lack the middle ear cavity. The bones that carry sound vibrations, instead of abutting on a tympanic membrane, connect with the jaw bones, and it is through these that the snake hears. The old saying "deaf as an adder," probably based on the absence of external ear openings, is erroneous.

Jacobson's Organ. This vomeronasal organ is small in turtles and crocodilians. It is highly developed in snakes and lizards and no longer connects to the nasal passages but opens directly into the mouth. The flickering, two-pronged tongue of the snake picks up chemical particles from the air. When the tongue is retracted, its tips are inserted into the openings of the pockets of Jacobson's organ. The sense of smell here plays a major role in the recognition of prey, enemies, or potential mates.

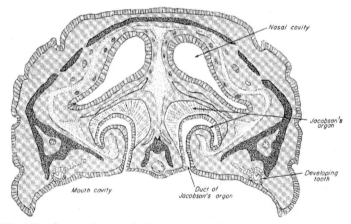

FIG. 3-8. Section through the head of a lizard, *Lacerta*, showing Jacobson's organ.

Collateral Reading and General Reference

Gadow, Hans. *Amphibia and Reptiles.* Cambridge Natural History, vol. 8. London, 1909. (An early work that contains a tremendous amount of fundamental information on the amphibians and reptiles.)

Reese, A. M. *The Alligator and Its Allies.* London: Putnam's, 1915. (The only attempt at a comprehensive survey of the features of this group.)

Romer, A. S. *Osteology of the Reptiles.* Chicago: University of Chicago Press, 1956. (An excellent, comprehensive account of the skeletal system of the reptiles, living and extinct.)

Smith, M. A. *The Fauna of British India. Reptilia and Amphibia.* Vols. 1–3. London: Taylor and Francis, 1931–1943. (A regional systematic study that includes a great deal of basic biology. Concise and clear accounts of structure.)

Weichert, C. K. *Anatomy of the Chordates.* 2d ed. New York: McGraw-Hill, 1958. (Or any other standard comparative anatomy text.)

Young, J. Z. *The Life of Vertebrates.* London: Oxford University Press, 1950.

ORIGIN AND EVOLUTION

OF AMPHIBIA

DURING THE LONG COURSE of evolutionary history groups occasionally arise that make major adaptive shifts. Because of structural
changes, which actually may be relatively minor, they are able to adopt
new modes of living and to move into environments that were closed to
their immediate ancestors. Surely one of the most important of such shifts
was that made by the amphibians, the first vertebrate animals to leave the
water and spread out over the land. Indeed they were the only vertebrates
to do so, for from them have come all the higher types, the reptiles, birds,
mammals, and man himself. An unusual interest, then, attaches to the early
evolutionary history of these relatively inconspicuous inhabitants of the
land.

SOME PALEONTOLOGICAL CONSIDERATIONS

Unfortunately, amphibians as a group do not leave very good fossils.
Many of them are small and have delicate skeletons, and their bones are
easily scattered or crushed. This is especially true for the more recent forms,
most of which are smaller than their early amphibian ancestors. And so
there are great gaps in our record of amphibian evolution. Some are being
slowly filled as paleontologists turn more of their attention to the tiny
scattered vertebrae that may be the only remnants of some once numerous
form. Some gaps will never be filled in. There will always be a certain
amount of guesswork involved in our attempts to construct amphibian
family trees. But the guesses of paleontologists are not blind stabs in the

54

dark. Through studies of forms that have left clear and long-continued records, certain evolutionary principles have been worked out to serve as guides. One of these is that there is a strong tendency toward irreversibility in evolution. An animal that has lost its legs does not later give rise to an animal possessing legs, a bone that has disappeared from the skull of an ancestral form will not reappear in the skulls of descendants. We may never be able to say definitely just which fossil forms were the forerunners of the modern salamanders, but we can be quite sure that they were not animals that had completely lost their legs. Thus by a process of elimination we may be able to narrow the field to one or a few probable ancestors even for those modern animals that have no fossil history.

It is sometimes assumed that the basic criterion in the classification of animals is how recently two forms have diverged from a common ancestor. Thus, if two species are put in the same genus, their common ancestor is thought to be closer to us in time than is the common ancestor of these two and a third species which is placed in a different genus. But this criterion is really quite limited in its applicability. The length of time two forms have been separated is only roughly correlated with the degree of

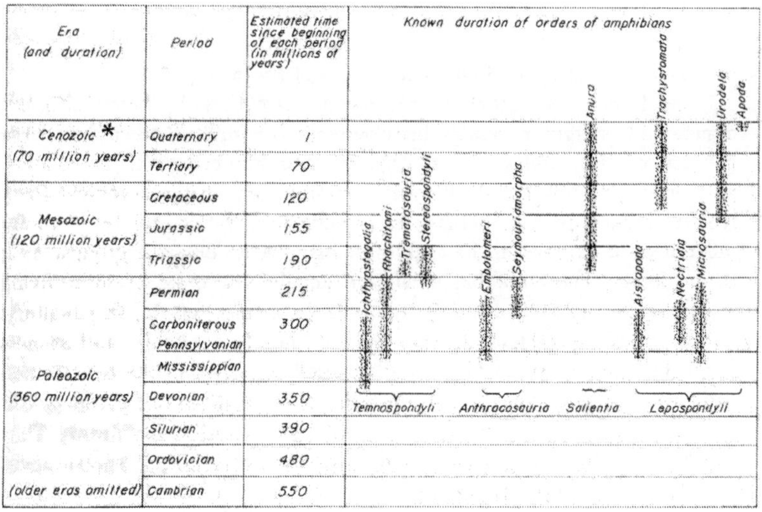

Era (and duration)	Period	Estimated time since beginning of each period (in millions of years)	Known duration of orders of amphibians
Cenozoic * (70 million years)	Quaternary	1	
	Tertiary	70	
Mesozoic (120 million years)	Cretaceous	120	
	Jurassic	155	
	Triassic	190	
Paleozoic (360 million years)	Permian	215	
	Carboniferous Pennsylvanian	300	
	Mississippian		
	Devonian	350	Temnospondyli Anthracosauria Salientia Lepospondyli
	Silurian	390	
	Ordovician	480	
(older eras omitted)	Cambrian	550	

Orders shown across the chart: Ichthyostegalia, Rhachitomi, Trematosauria, Stereospondyli, Anura, Seymouriamorpha, Embolomeri, Aistopoda, Nectridia, Trachystromata, Microsauria, Urodela, Apoda.

* The Tertiary is frequently divided into epochs. Beginning with the oldest, these are: Paleocene, Eocene, Oligocene, Miocene, and Pliocene. The Quaternary is also divided into Pleistocene and Recent.

FIG. 4-1. The geologic time scale, showing the distribution of the orders of amphibians in time.

divergence they show. Organisms evolve at different rates. Of two lines, one may evolve very slowly and remain close to the ancestral type, the other may evolve rapidly into something quite different. Very early in the evolutionary history of the amphibians, perhaps even before they were true amphibians, the stock apparently split into several groups. One gave rise to the modern salamanders, another to the amphibian ancestors of the reptiles and higher tetrapods, and another to the modern frogs. But because the frogs and salamanders retain a basically similar structural pattern and mode of life, we place them in the same class, whereas the reptiles, birds, and mammals, which have departed widely from the amphibian pattern, are placed in different classes. Degree of divergence, rather than closeness of descent, is thus the basic criterion of classification.

The introduction of the factor of time brings another complication into schemes of classification. Suppose at a given period we have two genera that are obviously closely related. Each of these genera gives rise to a line, the later members of which are so distinct as to be placed in different families. Should we classify the two original genera in the families to which they gave rise, a vertical classification, or should we put them together in a third family, a horizontal classification? Either would be technically correct. Which scheme we adopt will probably depend on the completeness of the fossil record connecting the various forms, a matter of chance. Most classifications are actually based on combinations of the two methods.

It should be stressed that any classification is not an objective reality but a man-made system, necessary because we find it difficult to study a large number of objects unless we can organize them into categories. If we knew all there is to know about all living organisms, and had a complete fossil record of every species that has ever lived, there would still be room for differences of opinion on the rank to be assigned to different groups. As it is, we know almost nothing about the detailed structure of most living species, and the fossil record is, and will necessarily remain, fragmentary. As we study more intensively the material already available, and as new forms, both living and extinct, are discovered, our ideas of the composition and relationships of the groups will shift. The classifications given in this and following chapters will almost certainly be revised in the future. They should not be taken as final or as indicating that everything is known about the relationships of the herptiles.

In studying the evolutionary history of a class it is possible to use either a horizontal or a vertical approach. With the former, we would consider all the amphibians of the Mississippian, then all those of the Pennsylvanian, and so on. This would give a clear picture of the fauna of any given period but would make it difficult to follow the various lines of descent. We shall here' adopt very largely a vertical approach, following one line through its evolutionary course, then going back and picking up another line. This method involves a certain amount of mental gymnastics, of jumping backward and forward in time. To do so it is necessary to keep the geological time scale clearly in mind. Figure 4-1 shows the time scale with the known duration of the amphibian orders for reference in the following discussion.

ORIGIN OF AMPHIBIA

The earliest known animal that can definitely be assigned to the amphibians (*Ichthyostega* of Greenland) is from freshwater beds of the late Upper Devonian. Hence it is among the fishes of the Devonian or earlier times that we must look for ancestors of the tetrapods.

The primitive vertebrates of the Silurian and early Devonian (class Placodermi) had lungs, bony dermal plates, and also bony internal skeletal structures. The sharklike fishes (class Chondrichthyes), which appeared relatively late in the Devonian, had lost the ability to produce bone so that their skeletons remained cartilaginous throughout life. Presumably they had also lost the lungs, which are not present in modern members of the class. The earliest amphibians had well-developed bones and probably lungs. Both because of their late appearance and because of their loss of bone, it is highly unlikely that the shark-like fishes could have been the precursors of the amphibians.

Ancestors of the modern bony fishes (Actinopterygii) were present in the early Devonian. The paired fins of these fish, supported only by cartilaginous rays, could hardly have developed into limbs able to bear the weight of an animal on land. The Actinopterygii too may be ruled out as ancestral to the tetrapods.

In the middle Devonian the Dipnoi appeared. This is the group to which the present-day lungfishes belong. Their fins were supported by bony elements, but in many respects the Dipnoi of the Devonian seem to have been already too specialized, especially in the structure of their teeth which were adapted to crushing shellfish, to have given rise to the amphibians.

Represented today by a single species, the coelacanth, the lobe-finned fishes (Crossopterygii) were abundant and diversified in the Devonian. They had lungs and their fins were supported by bony elements comparable to the bones of the tetrapod limb (Fig. 4-2). Furthermore, some of them developed true internal nares. The nasal sac of other fish is a sensory pouch on the side of the head, quite unconnected with the mouth. In some crossopterygians, a nasal passage led from the nasal sac to an opening in the roof of the mouth, bounded laterally by the bones of the upper jaw. This

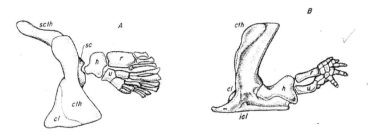

FIG. 4-2.   The shoulder girdle and pectoral fin of a crossopterygian (A); and
the same structures in an ancient, fossil amphibian (B) placed in a comparable
pose to show the basic similarity in limb pattern: h, r, and u, humerus, radius,
and ulna of the tetrapod and obvious homologues in the fish fin; cl, clavicle;
cth, cleithrum; icl, interclavicle; sc, scapula; scth, supracleithrum. [From
Romer; *The Vertebrate Body*, Saunders, Philadelphia, 1955.]

opening, the internal naris or choana, is a characteristic tetrapod structure.
While the nasal sac retains its sensory function, the passageway provides a
means by which air may be taken into the mouth and lungs when only the
external nostrils are above water. This probably increased the importance of
the lungs as respiratory organs.

The braincase of the crossopterygians was divided transversely into
anterior and posterior parts which apparently were slightly movable on each
other. No other vertebrates show this condition, and its functional significance is unknown. For a time it was thought to debar the crossopterygians
from the ancestry of the tetrapods. Recently, though, *Ichthyostega* has been
shown to retain traces of a transverse division of the braincase.

The crossopterygian notochord apparently remained large. The vertebra
consisted of a neural arch, with a small bone on either side just below the
arch, the pleurocentrum and, curving around the notochord ventrally, a
U-shaped bone, the intercentrum or hypocentrum. This rather loosely constructed vertebra was ill-adapted to withstand the stresses of life on land,
and it did not persist in later amphibians. The notochord is retained, if at
all, only as a series of cushioning pads between the vertebrae, and the
centrum tends to become a single bony unit fused to the neural arch.
Ichthyostega, however, still had vertebrae of the crossopterygian pattern.

Because of the structure of their fins and vertebrae, the presence of
choanae, and even the division of the braincase, the crossopterygians seem
well qualified to be the ancestors of the amphibians.

The question of why the ancestral amphibians should have deserted the
water to move out onto the land still remains unsolved. The answer possibly
lies in the climatic conditions of the Devonian when there were probably

periods of drought over wide areas. It must have happened many times that crossopterygians living in the fresh waters of semiarid regions were subjected to overcrowding as the waters shrank in the drying ponds. Animals able to heave themselves out onto the land, to push across it in search of larger bodies of water, or perhaps to augment the limited food supply available in the shrunken ponds by feeding on the already terrestrial insects and on the dying fish left stranded on the shores, may have had a decided advantage over more strictly aquatic forms. It has been argued that during a time of drought, conditions on land would have been quickly fatal to animals as susceptible to desiccation as the early amphibians must have been. On the other hand, when a drought is interrupted by rain, the ground and air become moist enough to permit cross-country migration by modern aquatic amphibians though the ponds may not fill up enough to relieve overcrowding. It may have been at some such time that the first step onto land was taken.

CLASSIFICATION OF AMPHIBIA

Very early in their history the amphibians split into several different groups, here classed as superorders. These superorders are characterized by different patterns of consolidation of the centrum (see Fig. 4-3). One line (superorder Temnospondyli), that began with an Ichthyostega-type vertebra, passed through a stage in which there was a double centrum composed of two wedge-shaped blocks—the pleurocentrum and hypocentrum. The pleurocentrum of the later temnospondyls was reduced and the hypocentrum became the definitive centrum. The superorder Anthracosauria also passed through a stage with a double centrum, consisting of two successive discs rather than wedges. In this line the pleurocentrum came to form the main body of the centrum and the hypocentrum was lost or persisted only as a vestige. This was the line that gave rise to the amniotes. The centra in the remaining two superorders (Salientia and Lepospondyli) cannot be traced definitely to the Ichthyostega-type vertebra. The centrum of the Lepospondyli is a simple, spool-shaped structure which apparently ossifies directly around the notochord without being laid down first as cartilage. That of the Salientia forms through chondrification and then ossification of part or all of the perichordal tube (sheath around the notochord) below each neural arch. Studies on the embryonic development of living members of these two groups indicate that their centra form from the same region as does the pleurocentrum of the amniotes, and their vertebrae are thus probably pleurocentral rather than hypocentral.

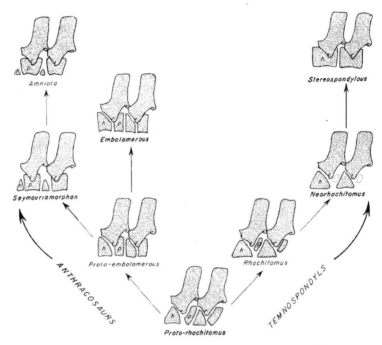

FIG. 4-3. The phylogeny of vertebral types: h, hypocentrum; p, pleurocentrum (true centrum). Viewed from left side. [After Romer.]

Superorder Temnospondyli

The Temnospondyli was a widespread and flourishing group during the later Paleozoic. The earliest members were aquatic but some of the later forms were well adapted to life on land. Then, in the late Permian and Triassic, the temnospondyls returned to life in the water, and some apparently even became marine. The line died out at the end of the Triassic.

Order Ichthyostegalia. This order includes the most primitive of all amphibians. *Ichthyostega* was a fairly large animal with a skull 150 mm. or more in length. It seems to have been intermediate between a fish and a tetrapod in appearance. It had short, stubby, pentadactyl limbs instead of fins, and a tail something like that of a modern lungfish with a caudal fin supported by fin rays. It had a short, rounded snout and a relatively long skull table (the posterior part of the skull roof). In these proportions it was intermediate between the crossopterygians and the later amphibians in which the snout region is well developed and the skull table reduced. Lateral line canals were present, suggesting that these aquatic sense organs were still important to *Ichthyostega*. The occipital condyle was single—another fish-like character. The internal nares were bounded laterally by a slender process from the maxilla rather than by a broad bar of bone as in most tetrapods. *Ichthyostega* must have spent most of its time in the water but may also have been able to crawl out on land.

Besides *Ichthyostega*, the order includes several other genera of Carboniferous amphibians. Some of them, at least, retained an armor of fish-like, bony, ventral scales, and also had traces of scales on the back and flank. Although very primitive in most respects, they all showed specializations indicating they were side branches rather than a part of the main stem of amphibian evolution. Thus one of the bones of the skull (the intertemporal), present in other primitive amphibians, was lacking in all the known genera of ichthyostegalians.

Order Rhachitomi. The rhachitomes of the Carboniferous and Permian were the most abundant of all fossil amphibians in number of genera. Almost all the large amphibians of the Permian belong in this order. The more primitive members of the group were probably largely aquatic but some of the later rhachitomes were apparently well adapted to life on land. Each neural arch had a double centrum composed of interlocking, wedge-shaped blocks, the typical rhachitomous vertebra. The occipital condyle was single, double, or triple. The intertemporal bone was present in early members of the group but disappeared in later forms.

The loxommids, a group of primitive forms from the Upper Mississippian and Pennsylvanian, are known only from skull elements. Whether or not they had rhachitomous vertebrae, their skulls were in many respects similar to what we should expect to find in the early rhachitomes. They were specialized, though, in having peculiar, large, elongate orbits. Probably they were an early side branch of the rhachitomous stock.

Unquestioned primitive rhachitomes were present in the Pennsylvanian and had developed considerable morphological variation. Some were very long-faced, others were short-faced. The latter may have been specialized or may simply have been larval or even neotenic forms.

The typical rhachitomes of the Permian represent the culmination of development of terrestrial life in the temnospondyls. Many of them were quite stout-limbed and seem to have been well adapted for life on land. Some were giants among the amphibians—*Eryops* of the Lower Permian was about 150 cm. long.

Order Trematosauria.   The trematosaurs were a spectacular group, probably derived from the later rhachitomes. They are known only from the Triassic. The triangular skull was rather high and narrow, sometimes with a very elongated snout. Grooves indicate the presence of lateral line sense organs. The intertemporal bone was absent, and the occipital condyles were double. The rest of the body is poorly known, but was apparently long and slender with feeble legs. The vertebrae were still basically rhachitomous but the pleurocentra were reduced and sometimes remained cartilaginous.

It is apparent that the trematosaurs were aquatic in habit and their sharp-pointed teeth suggest that they ate fish. In Greenland and Spitzbergen their fossils occur in beds with abundant fish remains. Since these beds are, in part, marine, it may be that the adults were adapted to enter the sea to feed but returned to fresh water to breed. There are indications that in the early Mesozoic the actinopterygian fishes started moving from fresh water to salt water and it may be that for a short time the trematosaurs went along a parallel line of evolution. For them the experiment was not a success, for the trematosaurs became extinct by the end of the Triassic.

Order Stereospondyli.   These last and most degenerate members of the Temnospondyli had given up their foothold on land and slipped back to an aquatic, mainly bottom-dwelling existence. The body was flattened and the limbs were too small and weak to have supported the animal on land. Lateral line canals were present on the skull. The intertemporal was absent, and the occipital condyles were double. The pleurocentra were reduced, seldom ossified, and often entirely absent. The hypocentra were always highly developed and sometimes formed complete discs around the notochord. The stereospondyls evolved from the rhachitomes in the Permian and were rather common in the Triassic.

The dividing line between the rhachitomes on the one hand and the stereospondyls on the other is not a sharp one. The group known as neo-rhachitomes might well be placed at the top of the former rather than at the bottom of the latter. Like all the stereospondyls, the neorhachitomes were characterized by features associated with a degenerative trend toward an aquatic existence. Their skulls were flattened and marked with lateral line grooves. Many of the elements of the braincase were unossified, or at least poorly developed. The limbs were small with poorly ossified bones and the shoulder girdles were much expanded and flattened ventrally. The neorhachitomes lived in the late Permian and early Triassic.

Obviously derived from the neorhachitomes, the long-snouted capitosaurs of the Triassic form a rather compact group showing little variation. Some are notable for their size. *Mastodonsaurus*, with a skull length of over 90 cm., was the largest of the amphibians.

The remaining stereospondylous amphibians of the Triassic are known as the short-faced stereospondyls. The skull was extremely broad and flat. The face was relatively short, but the part of the skull behind the eyes was decidedly elongated. The external nares lay close together and were much enlarged; it is possible that they accommodated the tips of the tusks of the lower jaw. Like the capitosaurs, the short-faced stereospondyls apparently evolved from the neorhachitomes and died out at the end of the Triassic.

Superorder Lepospondyli

A group of small amphibians which flourished in pools of the Carboniferous had the centrum of the vertebra formed as a bony cylinder around the notochord instead of as separate hypocentral and pleurocentral blocks. They were all animals of modest size; many of them had lost their limbs and were eel-like in appearance. These Paleozoic lepospondyls are divided into three orders: Aistopoda, Nectridia, and Microsauria. Three modern orders of amphibians (Trachystomata, Caudata, and Apoda) have vertebrae apparently constructed on the same basic plan. Although there are great gaps in the fossil record between the earliest known members of these orders and the lepospondyls of the Paleozoic, they seem to be members of the same stock and are included in the same superorder.

Order Aistopoda. This order includes only a small number of elongate, limbless amphibians. Some are snake-like in build and have more than a hundred vertebrae. They early deserted the amphibian trend toward terrestrial life, if indeed their ancestors ever followed that course, and were all aquatic. They appeared in the Mississippian and died out at the end of the Pennsylvanian.

Order Nectridia. There is some question about whether the aistopods and nectridians should be placed in different orders since they show many similarities. The nectridians were somewhat more varied, but all had the neural and haemal arches of the caudal vertebrae expanded into fanlike structures. Very common in the Pennsylvanian pools were a number of slender, eel-shaped little nectridians with long, pointed heads and with the limbs reduced or absent. Also abundant in the Pennsylvanian, and lasting into the Lower Permian, was another group of nectridians. They were somewhat larger (some reached lengths of about 600 mm.) and had flattened bodies, reduced limbs, and weird-looking heads in which the posterior corners of the skull were drawn back into a pair of grotesque horns. They were probably bottom dwellers, and swam by skate-like undulations of the body.

Order Trachystomata. This order includes two modern genera, *Siren* and *Pseudobranchus,* and a number of extinct forms, known only from vertebrae, which go back to the Cretaceous. They are usually classified with

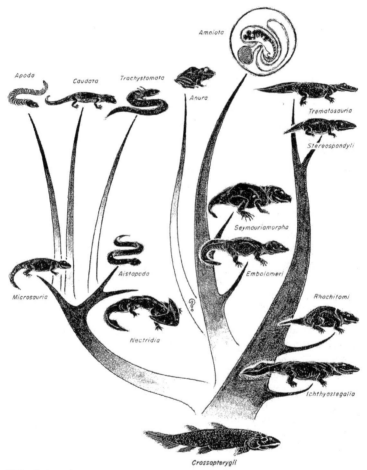

FIG. 4-4. The phylogenetic arrangement of the orders of amphibians.

the Caudata, but they differ sharply from all known salamanders and, especially in the detailed structure of their vertebrae, show marked resemblances to the aistopod-nectridian stock. They, too, are eel-shaped and have expanded arches on the tail vertebrae. The hind limbs are absent, the fore ones minute. (Since the pectoral girdle is not attached to the rest of the skeleton, the tiny limb bones usually become separated from the rest of the body shortly after death. Limbs are not known for any fossil trachystomes, but their presence in the modern forms shows that front legs must also have been present in the ancestral species and raises the question of whether some of the supposedly legless aistopods may not also have had vestigial forelimbs.)

Order Microsauria. The microsaurs were small creatures which in general looked more like modern salamanders than did any of the lepospondyls previously mentioned. They varied considerably in structure; some paralleled the aistopods and nectridians in the loss of limbs but most seem to have had a more normal tetrapod build. Together with the aistopods and nectridians, they flourished during the Carboniferous; indeed, the lepospondyls seem to have been the most abundant amphibians in the swamp pools of that time. Microsaurs became extinct during the Permian.

Order Apoda. The caecilians are slender, worm-like creatures, practically blind and without limbs or limb girdles. Their compact skulls are modified for burrowing and lack some of the bones present in primitive amphibians. No fossil caecilians have ever been found to help bridge the gap between the modern apodans and their possible Paleozoic ancestors. But their rather striking resemblance to *Lysorophus*, an aberrant, worm-like microsaur of the Permian, strongly suggests that the caecilians descended from this or some similar microsaur rather than from the more specialized aistopod or nectridian lines.

Order Caudata. Modern salamanders are mostly small animals with long slim bodies and tails, and rather feeble legs. Like the other modern amphibians, they show a strong tendency toward a reduction in the bones of the skull. They have been traced as far back as the Jurassic, but there is still an enormous gap in the record between them and the lepospondyls of the Paleozoic. The aistopods and nectridians, though, were even then very specialized; it is almost certain that we should look for the ancestors of the Caudata among the microsaurs.

The scattered salamander vertebrae from the Jurassic and Lower Cretaceous do not seem to represent any of the modern families. Vertebrae from the Upper Cretaceous have been described as belonging to an extinct family, Scapherpetonidae. Another family, Batrachosauroididae, has been erected for a single genus known only from the Miocene. Other fossil salamanders have all been assigned to modern families (for descriptions of these families, see Chap. 12). Of these, the Hynobiidae, Ambystomatidae, Salamandridae, and Plethodontidae go back to the Paleocene and the Proteidae to the Eocene.

Cryptobranchidae were present in the Miocene. One large species, *Andrias scheuchzeri*, was described in 1726 as the mortal remains of a human sinner drowned by the Noachian Deluge, and named "Homo diluvii testis" (man witness of the flood). This, of course, was in the days before Darwin, when it was popular to explain away all fossils as having been left over from the Flood.

The Amphiumidae have so far been traced no further back than the late Pleistocene.

Superorder Salientia

The frogs, with their long hind legs and shortened backbones, are among the most specialized of all vertebrates. They cannot be traced back to any known group of Paleozoic amphibians, but evidence that their vertebrae are pleurocentral suggests that they may be allied to the anthracosaurs.

For a time it was thought that *Protobatrachus* of the Lower Triassic represented an intermediate stage in the evolution of the frogs. The single specimen known is that of an animal with a frog-like skull, a long tail, and long hind legs. Its pelvic girdle was not attached to a sacral vertebra, a structural weakness that would have prevented the animal from using its legs for jumping although they were apparently modified for that purpose. It is highly probable that the single known specimen of *Protobatrachus* was not a true evolutionary intermediate but a metamorphosing tadpole in which the tail had not yet been resorbed nor the sacral connection formed. But until we know what the adult was like, we cannot assign it a definite place in the classification of the frogs.

Order Anura. A series of unusually well-preserved fossils from the mid-Jurassic of Patagonia show that animals then lived that were undoubtedly frogs in the modern sense. They were rather large for frogs, about 130 mm. in length, and had ribs and relatively short legs. The family Notobatrachidae has been erected for this very primitive species. Other more fragmentary, scattered remains from the Upper Jurassic of Europe and North America have been placed in the family Montsechobatrachidae. One other family of

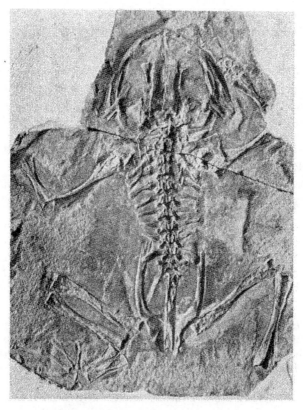

FIG. 4-5. *Notobatrachus degiustoi* Reig, a primitive frog from the Jurassic of Patagonia. [Courtesy Professor Osvaldo Reig.]

frogs is entirely fossil. This is the Paleobatrachidae, known from the Eocene and Miocene.

The remaining fossil frogs have all been placed in families that have living relatives (for descriptions of these families, see Chap. 13). The differentiation of many of these modern families seems to have taken place rather early in the evolution of the frogs. The Discoglossidae may have been present in the Upper Jurassic, and the Leptodactylidae and Pipidae in the Cretaceous. By Eocene times the Ranidae, Bufonidae, Pelobatidae, and Pelodytidae had appeared. The Hylidae are known from the Oligocene. The other families have not yet been reported from the Tertiary, but this is undoubtedly because of the incompleteness of our knowledge. It may someday be shown that most frog families go back to Mesozoic times.

Superorder Anthracosauria

This superorder includes the members of that line of amphibian descent in which the pleurocentrum came to form the main body of the centrum and the hypocentrum was progressively reduced. Anthracosaurs appeared first in the Mississippian and, as amphibians, died out at the end of the Paleozoic. But before they disappeared, they gave rise to true terrestrial vertebrates, the reptiles.

Order Embolomeri. These were common amphibians of the Carboniferous, primitive in many ways, yet in some respects already foreshadowing the reptiles. Both the intercentrum and the pleurocentrum were complete discs. There was a single occipital condyle and the intertemporal bone was still present. The later forms, at least, had long bodies, powerful tails, and small limbs, and some had very elongated snouts. They were almost certainly aquatic, and probably fish eaters. Some were very large; *Pteroplax* reached an estimated length of more than 450 cm. The embolomeres died out in the Lower Permian.

Order Seymouriamorpha. The seymouriamorphs flourished from the Upper Pennsylvanian to the Upper Permian. They were moderate in size, stockily built, with stout limbs, and were certainly capable of walking on land. The group takes its name from what is probably the most widely known fossil amphibian, *Seymouria.*

None of the well known seymouriamorphs was generalized enough to be considered a direct ancestor of the reptiles; apparently that ancestor was some as yet unknown or poorly known member of the order. A number of reptilian characters were present in the group. Among these were five digits in the front foot, swollen neural arches, and a true pleurocentrum in combination with a reduced, crescent-shaped intercentrum. Against them may be opposed such nonreptilian characters as an intertemporal bone in the skull, traces of lateral line grooves, the presence of more bones in the lower jaw than are found in reptiles, and a single sacral vertebra.

So nearly is *Seymouria* intermediate in structure between the amphibians and reptiles that it has sometimes been classified as the most advanced of the former, sometimes as the most primitive of the latter. But some of the fossils assigned to the group seem to represent aquatic larval stages. This strongly suggests that the seymouriamorphs did not lay shelled, amniote eggs on land, and are probably better included with the amphibians.

Collateral Reading and General Reference

Case, E. C. "A census of the determinable genera of the Stegocephalia." *Transactions of the American Philosophical Society*, new series, vol. 35, pt. 4, pp. 325–420, 1946. (The name "Stegocephalia" was formerly used for an order —or suborder—comprising all the pre-Jurassic amphibians. Later work showed that these amphibians belonged to several different phyletic lines no more similar to each other than are the various modern orders. The term no longer has any precise classificatory significance, but it is one that the student will encounter in the literature and with which he should be familiar. Case gives concise summaries of the characters of the families and genera. His arrangement of the higher categories is quite different from that of Romer.)

Piveteau, Jean. *Traité de Paléontologie.* Tome V. *Amphibiens, Reptiles, Oiseaux.* Paris: Masson et Cie., 1955. (A well illustrated, modern treatment of the fossil amphibians and reptiles.)

Romer, A. S. *Vertebrate Paleontology.* 2d ed. University of Chicago Press, 1945. (Includes a good summary of amphibian evolution, less technical than the other references given here.)
"Review of the Labyrinthodontia." *Bulletin of the Museum of Comparative Zoology at Harvard University*, vol. 99, no. 1, 1947. (An excellent, comprehensive survey of the Temnospondyli and Anthracosauria, which Romer treats as orders of the superorder Labyrinthodontia. Since it is probable that the Salientia, and perhaps also the Lepospondyli, are closer to the anthracosaurs than the latter are to the temnospondyls, we have reluctantly dropped the time-honored term Labyrinthodontia from our classification.)

Watson, D. M. S. "The evolution and origin of the Amphibia." *Philosophical Transactions of the Royal Society of London* (B), vol. 215, 1926. (In this and other papers Watson laid the foundation for the modern classification of amphibians.)

Williams, E. E. "Gadow's Arcualia and the Development of Tetrapod Vertebrae." *Quarterly Review of Biology*, vol. 34, no. 1, pp. 1–32, 1959. (The original work on which we have based most of our classification of the higher categories.)

CHAPTER

ORIGIN AND EVOLUTION
OF REPTILIA

THE AMPHIBIANS were like the early explorers of the New World who still called the Old World home. They invaded and explored the land but most of them could not travel far from their aquatic homeland and necessarily returned to carry on the reproduction of their kind. The reptiles were the true colonizers who settled down to live and reproduce in the New World of dry land.

THE CLEIDOIC EGG

The development of the cleidoic egg marked the dividing line between the amphibians and the reptiles. Many present-day amphibians (caecilians, a few salamanders, a number of frogs) lay their eggs away from water (see Chap. 6). Some produce eggs that are relatively large-yolked and the young, instead of going through a larval stage, hatch in the adult body form. Such eggs need only shells and extra-embryonic membranes to make them cleidoic. Most of the amphibians that lay terrestrial eggs live in mountainous regions where there are few large open bodies of quiet water to which they may resort for breeding. It is possible that the reptilian egg evolved as an adaptation to mountainous conditions. At any rate, the first animals that had reached the morphological level of true reptiles appeared in the late Pennsylvanian, a time of active mountain building.

70

THE REPTILE SKULL

The seymouriamorph skull was broad and solidly roofed over by dermal bones. These bones were arranged in three rows on each side: a median row including the frontal and parietal; a lateral row with the postorbital and squamosal; and a marginal row in which the jugal and quadrate were major elements. The brain was very small and encased in a narrow, bony box, the braincase, lying beneath the center of the skull roof. The muscles of the jaw attached to the underside of the roofing bones. During the evolution of the reptile skull, openings developed between the bones of the once-solid skull roof. If the openings developed along the sutures between bones of the median and lateral rows, a pair of dorsal temporal openings formed; if between the bones of the lateral and marginal rows, a pair of lateral temporal openings appeared. In some reptiles both dorsal and lateral openings developed (see Fig. 5-1).

The subclasses into which the reptiles are divided are defined, in part, by the nature of the temporal openings in the skull. Early workers who used this character paid most attention to the arches, the bars of bone lying below the openings. This is why the names of the types of skulls and of some of the suborders are derived from the Greek root "apse," which means arch. Thus the diapsids are reptiles with two arches, hence two pairs of temporal openings in the skull.

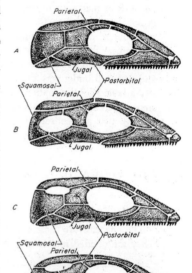

FIG. 5-1.
Diagrammatic view of reptilian skulls to show variation in temporal openings. A, The primitive type without openings, anapsid; B, the type found in the ancestors of mammals, with a lateral opening, synapsid; C, the type found in plesiosaurs, etc. with an upper temporal opening, euryapsid; D, the type found in *Sphenodon*, the crocodilians and other archosaurs, and in modified form in the lizards and snakes, diapsid.

RADIATION OF THE REPTILES

The reptiles of the present are but a remnant of the hordes that roamed the world during earlier geologic times. Once they had solved the problem of life away from the water, reptiles began a rapid radiation which enabled them to take advantage of all the new environments thus opened to them. They faced no competition from higher groups. They were better adapted to terrestrial conditions than were the amphibians and there were as yet no birds or mammals to dispute their possession of the land. They spread rapidly. Some took to the air, others returned to the water, though with a curious reversal of the amphibian pattern, for the marine reptiles either had to bear their young alive or return to land to lay their eggs. This rapid radiation began soon after the origin of the group in the Pennsylvanian and was well under way by the end of the Paleozoic. The enormous diversity attained by the reptiles in their heyday, and the reduction they have since suffered, is shown in a recent classification, which lists more than 170 families that are entirely fossil. This is nearly four times the number of families that have living members.

CLASSIFICATION OF REPTILES

Subclass Anapsida

The anapsids are those reptiles with no temporal openings in the skull. They are divided into two orders.

Era (and duration)	Period	Known duration of orders of reptilia							
Cenozoic * (70 million years)	Quaternary								
	Tertiary								
Mesozoic (120 million years)	Cretaceous								
	Jurassic								
	Triassic								
Paleozoic (360 million years)	Permian								
	Carboniferous Pennsylvanian	Anapsida	Lepidosauria	Archosauria	incertae sedis	Ichthyosauria	Euryapsida	Synapsida	
(older periods and eras omitted)	Mississippian								

Diagonal order labels across the chart: Testudinata, Rhynchocephalia, Squamata, Crocodilia, Eosuchia, Saurischia, Ornithischia, Pterosauria, Ichthyosauria, Sauropterygia, Cotylosauria, Thecodontia, Mesosauria, Protosauria, Pelycosauria, Therapsida.

    * The Tertiary is frequently divided into epochs. Beginning with the oldest, these are: Paleocene, Eocene, Oligocene, Miocene, and Pliocene. The Quaternary is also divided into Pleistocene and Recent.

FIG. 5-2. The geologic time scale, showing the distribution of the orders of reptiles in time.

FIG. 5-3. Anapsid reptiles: lower, *Limnoscelis*, a primitive, fish-eating cotylosaur; middle, *Bradysaurus*, a slow-moving, herbivorous cotylosaur; upper,
Macroclemys, the modern Alligator Snapping Turtle.

Order Cotylosauria. This order includes the primitive stem reptiles
which arose in the Pennsylvanian from a form perhaps like *Seymouria* and
reached a climax in the late Paleozoic. They were mostly squat, ungainly,
lizard-like animals with sprawling legs. They seem to have been semiaquatic,
suggesting that the amniote land egg may have developed before the adult
reptile had become fully terrestrial.

Some of the later forms (pareiasaurs) had limbs that had rotated inward,
enabling them to support the weight of a heavy body more efficiently than
could the sideward sprawling limbs of the amphibians and earliest cotylosaurs. The pareiasaurs were the largest of the stem reptiles, reaching
lengths of more than 300 cm. They were apparently slow-moving herbivores,
perhaps something like the modern hippopotamus in habits.

The cotylosaurs were small-brained creatures, and before the end of the
Triassic they had become extinct. But they left as descendants not only the
second order of anapsids, the turtles, but all later reptiles.

Order Testudinata. The turtles, although they belong to the most primitive group of reptiles, have persisted until modern times. Early in their
history they performed a feat that has never been accomplished by any
other group of vertebrates. Somehow they got their ribs outside their girdles so that the ribs could enter into the formation of a protective shell

without hindering the freedom of movement of the limbs. This proved to be a successful adaptation, for turtles have continued their placid existence ever since. While the dinosaurs flourished and then disappeared, while the grasslands spread during the Cenozoic and the mammals took over the face of the earth, the turtle, with its ribs still outside of its girdles, lived on practically unchanged and seemingly flourishes today as well as in the past.

Turtles have heavy skeletons in proportion to their size and they also have hard shells. Furthermore, they frequently live in aquatic habitats where dead animals may sink to the bottom and be covered with silt, thus being protected from the forces of decay and erosion until they become fossilized. For these reasons the turtles have left a better fossil record than have the lizards and snakes with their more fragile skeletons and terrestrial habits.

One suborder of the Testudinata, the Amphichelydia, is entirely extinct. When these most primitive of the turtles first appeared in the Upper Triassic, they already had the typical turtle armor fully developed. Apparently the head could not be pulled back into the shell. Most of the Amphichelydia had died out by the end of the Mesozoic, but one form, *Meiolania*, persisted in Australia until the Pleistocene. It must have been a huge beast, for it had a skull more than 60 cm. wide.

Modern turtles are divided into two suborders: Pleurodira, the side-necked turtles, which turn the neck sideways to tuck the head under the shell; and Cryptodira, which bend the neck in a vertical sigmoid curve to withdraw the head.

Most of the families of living cryptodires were present in the Cretaceous. The Dermatemydidae mainly includes rather primitive turtles of the Cretaceous and early Tertiary, but a single species, *Dermatemys mawi*, still exists in Central America. The Cheloniidae, Carettochelyidae, and Trionychidae were all present in the Upper Cretaceous. The Testudinidae may have also been present at that time and certainly were in the Paleocene. The Dermochelyidae are known from the Eocene, and the Chelydridae from the Oligocene.

There are only two families of pleurodire turtles. The Pelomedusidae go back to the Lower Cretaceous, but the Chelidae are not known earlier than the Pliocene. (For descriptions of the modern families of turtles, see Chap. 14.)

Subclass Lepidosauria

This is the first subclass of diapsid reptiles, those that have two temporal openings on each side. In the later lepidosaurs, one or both of the arches may be lost so that the nature of the openings is obscured, but these groups can be traced back to forms that were diapsid. The lepidosaurs apparently branched off from the cotylosaur stock some time in the Permian. Of the three orders in this subclass, two survive to the present day.

Order Eosuchia. These were rather small reptiles of the Permian and early Mesozoic. They were lizard-like in general appearance and perhaps also in behavior, but they differed from modern lizards in having the arch of bone that closes the lateral opening on the lower side complete. In true lizards this arch has disappeared. Most of the eosuchians had died out by mid-Triassic times. However, *Champosaurus*, a long-snouted, aquatic diapsid reptile rather like a modern gharial in build, which lived in the late Cretaceous and early Tertiary, may have been a late-surviving member of the eosuchian stock.

Order Rhynchocephalia. The beak-headed reptiles resemble the eosuchians but differ from them in having the teeth fused to the edge of the jaw rather than set in sockets, and in having an overhanging beak on the upper jaw, from which they get their name. Like the eosuchians, and

FIG. 5-4. Lepidosaurian reptiles: upper left, Tuatara, *Sphenodon;* upper right, Chameleon, *Chamaeleo;* lower left, Mosasaur, *Clidastes;* lower right, Rock Python, *Python.*

unlike the Squamata, they retain the two complete temporal arches. The rhynchocephalians evolved from the eosuchians about the beginning of the Triassic. One species, *Sphenodon punctatus,* still lives today on a few islands off the coast of New Zealand (see Chap. 17). This "living fossil" is among the most venerable of the vertebrates, for it appears to have survived with little change from the early Mesozoic. All other known rhynchocephalians are from the Triassic or Jurassic.

Order Squamata. This order includes the lizards and snakes, the really successful modern reptiles. They differ from the other lepidosaurs in the reduction of the bony framework on the side of the head by the loss of one or both of the arches. There are two suborders, the Lacertilia (lizards) and the Serpentes (snakes).

The dinosaurs were so spectacular that the study of them dominated paleontology for years. Twenty-five families of dinosaurs are currently recognized. In contrast, only five exclusively fossil families of lizards, and two of snakes (one may be an eell) are known. Now that the major outlines of dinosaur evolution have been filled in, paleontologists are turning their attention to the lesser forms that managed somehow to survive the destruction that overwhelmed the dinosaurs and hang on to the present day. Recently developed techniques of micropaleontology have stimulated the study of the tiny and fragile lizard jaws and snake vertebrae that tell the story of these little cousins of the dinosaurs. Enough has been done already to show that our limited knowledge of the fossil history of the Squamata results not from their lack of an extended evolutionary past nor their failure to leave a fossil record, but from our failure to look for the evidence. We may expect in the next few years an enormous increase in our knowledge of fossil snakes and lizards which will undoubtedly lead to some realignment of the families. Especially we may hope that a light will be thrown on the murky picture of the relationships of the snakes now lumped together in the family Colubridae—perhaps the most pressing systematic problem in herpetology.

The lacertilians probably branched off from the eosuchians some time in the Triassic, but the oldest known definitive lizards are from the Upper Jurassic. A separate family, Ardeosauridae, has been erected for these primitive forms. One other poorly known Upper Jurassic species has been doubtfully allocated to the family Xenosauridae, otherwise known only from Recent material. In the Upper Cretaceous, the rather primitive modern families Iguanidae and Agamidae appeared and also the more highly specialized Varanidae and Anguinidae. Fragmentary remains from the Upper Cretaceous and Eocene have been referred to the specialized Chamaeleonidae, but they may well be agamids instead. Some Paleocene and Eocene forms may be related to the modern Xenosauridae. The Typhlopidae, Gekkonidae, Lacertidae, and Amphisbaenidae, and perhaps the Cordylidae were present in the Eocene. The Teiidae and Helodermatidae appeared in the Oligocene, the Scincidae in the Pliocene. The other families are unknown as fossils. (For descriptions of the modern lizard families, see Chap. 15.)

The most spectacular lizards that ever lived were the mososaurs, huge, sea-going, fish-eaters with long heads, long tails, and paddle-shaped limbs. The largest reached 12 m. in length. They were common in all seas during the Upper Cretaceous but did not survive into the Cenozoic.

Our knowledge of fossil snakes is even more limited than is our knowledge of fossil lizards. The bones of the snake skull are loosely connected and after death are soon scattered and lost. Usually only the vertebrae are preserved in identifiable shape as fossils. The poverty of our knowledge is illustrated by the fact that out of nearly 300 genera included in the family Colubridae, about a dozen are known as fossils. Only for the large boids do we have anything even approaching an adequate record.

The snakes seem to have evolved from the lizards, probably toward the end of the Mesozoic, for our first record of a snake is from the Upper Cretaceous (extinct family Dirrilysiidae). Of the living families, the Boidae may also have been present in the Upper Cretaceous and were certainly represented in the Eocene. The more advanced Colubridae, Elapidae, and Viperidae are known from the Miocene. The other families are unknown as fossils. The characters that distinguish the modern families become much less definite as we trace them back to the early Cenozoic. It is probable that the radiation of the snakes is a recent phenomenon, and that some of the families were not present before the late Cenozoic. (For descriptions of recent families of snakes, see Chap. 16.)

Subclass Archosauria

As the name implies (*archon* = ruling, *sauria* = reptiles) the members of the second subclass of diapsid reptiles were the ruling reptiles of the Mesozoic. In addition to the dinosaurs, the group includes such varied types as the crocodilians, the flying reptiles (pterosaurs), the crocodile-like phytosaurs, and the ancestors of the modern birds. Although the end forms are so diversified, the archosaurs seem to be a natural group which may have branched off from the primitive eosuchians some time in the Permian. While they were of little importance in the evolution of our modern reptiles, they were the lords of the earth during the long Mesozoic era.

Order Thecodontia. This order includes the primitive archosaurian stock of the Triassic, the ancestors of the crocodilians, pterosaurs, dinosaurs, and also of the birds. Most of them were active little carnivores, rather lizard-like in build, with the hind legs always longer than the front legs. This is an indication of the trend toward bipedal locomotion which is so strongly marked in many of the later archosaurs.

Also included in the Thecodontia are the specialized Phytosauria, common in the Upper Triassic. With long bodies, long tails, and long jaws lined with many sharp teeth, the fish-eating phytosaurs superficially resembled their contemporary relatives, the early crocodiles, but differed in one apparently vital respect. Instead of having the nostrils at the tip of the snout, with the nasal chamber separated from the mouth by a secondary palate, the phytosaurs had the nasal openings located on top of the head between the eyes and leading directly into the mouth. Phytosaurs were less successful than the crocodiles for by Jurassic times they had disappeared.

Order Crocodilia. The crocodilians, which appeared first in the Upper Triassic, are among the most conservative of the archosaurs. The more specialized pterosaurs and dinosaurs, although successful for a time, died out by the end of the Mesozoic, but the crocodiles survived. Today they are the only living remnants of the once widespread and dominant archosaurian stock.

Modern crocodiles belong to two families: the Crocodylidae go back to the Upper Cretaceous; the Gavialidae were certainly present in the Pliocene and may have existed as far back as the Eocene. (For descriptions of the modern families of crocodiles, see Chap. 17.) In addition, eleven extinct families are known, including the marine thalattosuchians with fish-like tail fins and paddle-shaped limbs.

Order Pterosauria. These were archosaurs that took to the air. They represent a different line from the one that gave rise to the birds and their flight mechanism was more like that of the bats. They never developed feathers but depended entirely on skin membranes stretched from the extended tips of the fourth fingers to the ankles. Their wings were long and narrow and from what we know of their habitat and manner of living it is fairly safe to judge that they were soarers rather than strong flyers. Their small breast bones would not have permitted the attachment of muscles large enough to make them active flyers. They probably lived on cliffs near the sea coasts, feeding on fish, and depending on rising air currents for sailing and gliding. Some were apparently little larger than sparrows, but *Pteranodon*, the giant of the group, had a wingspread of about 8 meters. They appeared first in the Lower Jurassic and lived through the Cretaceous.

FIG. 5-5. Archosaurian reptiles: upper left, Triassic thecodont, *Saltoposuchus;* upper right, a primitive pterosaur; center, *Alligator;* lower left, a saurischian dinosaur, *Tyrannosaurus;* lower right, an ornithischian dinosaur, *Stegosaurus.*

The remaining two orders of archosaurs are collectively known as the dinosaurs. They were the most impressive animals that ever lived. Only through a knowledge of them can we grasp the full evolutionary capabilities of the reptilian type of body organization and mode of life. If it were not for the dinosaurs, we might conceivably theorize that reptiles, because of their less efficient methods of heat regulation, are unable to attain the body sizes reached by the larger mammals.

Order Saurischia. The two orders of dinosaurs can be distinguished by the structure of the pelvic girdle. The Saurischia had a triradiate (three-branched) pelvis similar to that found in many thecodonts. The pubic bone of the Ornithischia had swung back and down to lie parallel to the ischium and had developed a broad, new, forward projection, thus forming a quadriradiate (four-branched) pelvis.

The saurischians evolved from the thecodonts in the Triassic. They are divided into two groups: the carnivorous, bipedal dinosaurs (theropods), and the herbivorous, quadrupedal dinosaurs (sauropods). The theropods included a number of large, heavily built flesh-eaters, the biggest of·which, was the huge and spectacular king of the carnivores, *Tyrannosaurus rex.* This creature was about 15 m. in length and stood about 6 m. high. It was undoubtedly the most dangerous carnivore the world has ever known. Other theropods were small, lightly built, predaceous animals, some not much larger than a barnyard chicken. Also classified with the theropods are the Triassic ancestors of the sauropods which had longer and more stoutly built front legs than the other theropods.

True sauropods appeared in the Jurassic. They had reverted to quadrupedal locomotion, although their front legs were still usually much shorter than their hind legs. They were long-necked, long-tailed, and small-headed. *Brachiosaurus*, the largest, though not the longest, reached about 24 m. in length and may have weighed nearly 50 tons. Though they were the largest animals that ever roamed the earth, they were rather defenseless creatures, relying on their size and habits to protect themselves from the carnivorous theropods. They probably lived in marshes, grazing among the succulent plants much as cows do nowadays. The bipedal *Tyrannosaurus* would have had heavy going in such terrain.

Order Ornithischia. The ornithischian dinosaurs differed from the saurischians in having a quadriradiate pelvis. Some were bipedal, though not as fully as most of the theropods; many lines reverted to a quadrupedal gait. They were all herbivorous. Ornithischians were late arrivals among the reptiles. They were very rare before the Upper Jurassic, but they later underwent an extensive radiation before their extinction at the close of the Cretaceous.

Such bizarre types as the armored dinosaurs, the horned dinosaurs, the aquatic duckbills, and the crested dinosaurs appeared among the ornithischians. The horned dinosaurs were rather large, though they never approached the size of the largest saurischians. In bulk and appearance an animal such as *Triceratops* must have been something like a modern rhino, with horns on the head and a bony shield projecting back from the posterior margin of the cranium over the neck and shoulder region. This was probably a rather efficient protection against being nipped on the back of the neck by *Tyrannosaurus.*

The armored dinosaurs were weird-looking creatures. Many had spectacular bony plates along the nape of the neck, the back, and the dorsal margin of the tail. Others had bony plates all over the back and looked like huge tortoises. Some developed large spines and nobs along the sides of the body and on the tip of the tail, which they probably swung in defense much as a'modern alligator does. A tail armed with such large bony spines was undoubtedly a very effective weapon.

The duckbill dinosaurs were aquatic. They probably fed like modern ducks, wading around the edges of marshes and reaching down to sieve out food with their large, flat, duck-like bills. Crested dinosaurs developed bony protuberances on top of their heads. They were cavernous structures, the internal chambers being continuous with the nasal air passages. These may have functioned as air reservoirs to permit the animals to remain submerged for long periods of time.

Subclass Uncertain

Order Mesosauria. This order was erected for a single genus of reptiles (*Mesosaurus*) found in early Permian beds on either side of the South Atlantic. *Mesosaurus* had a lateral temporal opening as did the Synapsida discussed below, but the opening was placed far down on the cheek and was almost certainly evolved independently. In other ways *Mesosaurus* differed greatly from the synapsids, nor does it seem to have been related to any of the other main reptilian stocks. It may have represented an early, short-lived side branch of the Carboniferous stem reptiles. It was less than a meter in length, and had a long snout armed with many sharp teeth, a long and powerful tail, and well-developed hind legs. It probably lived in fresh water and preyed on fish.

Subclass Ichthyopterygia

The Ichthyopterygia were reptiles with a dorsal temporal opening. They were a compact group, highly modified for marine life and showing little resemblance to any of the other reptilian stocks. The subclass contains only one order.

Order Ichthyosauria. The fish lizards were the dominant marine reptiles of the Mesozoic seas. They were built superficially like streamlined fish and were about the size of small porpoises. The eyes were large and the nostrils were set just anterior to them; the head joined the body with no perceptible neck. The swimming organ was a fish-like tail. The long jaws, armed with many sharp teeth, indicate that the ichthyosaurs were fish eaters. These animals probably occupied the same ecologic niche as the porpoises do today. They were obviously no better adapted for coming out on land to lay eggs than are the modern whales and it was long suspected that they bore their young alive. This was confirmed by the spectacular discovery of fossils

FIG. 5-6. Marine reptiles: upper, a plesiosaur; middle, an ichthyosaur; lower, *Mesosaurus.*

that show traces of well-developed embryos within the body of the mother. Ichthyosaurs appeared in the early Triassic and survived until the Upper Cretaceous.

Subclass Euryapsida

These reptiles, like the ichthyosaurs, had a dorsal temporal opening, and the later forms were also marine. But their aquatic specializations were entirely different, and there is little doubt that the two subclasses represent independent lines of reptilian evolution. There are two orders of Euryapsida.

Order Protorosauria. This order includes some primitive, rather lizard-like animals of the Permian and Triassic, mostly lightly built little creatures with rather long, slender limbs. A grotesque member was the Triassic *Tanystropheus;* it was about 75 cm. in length, and had a neck nearly as long as the rest of the body and tail combined. The lengthening of the neck resulted from the elongation of the individual vertebrae. When the first complete skeleton was discovered, it was found that the anterior part of the body had previously been described as belonging to a flying reptile, while the trunk had been thought to be that of a small dinosaur. No one has yet made what seems to be a sensible suggestion as to its mode of life.

Order Sauropterygia. The dominant sauropterygians were the marine plesiosaurs. Perhaps the most spectacular reptiles ever to invade the seas, they have been likened to a snake strung on the body of a turtle. The head was small, the neck long with many normal-sized vertebrae (76 in one form), rather than the elongated vertebrae characteristic of the protorosaurs. The body was short and flattened, the limbs large and heavy, the tail long. Some attained lengths of 15 meters. Presumably the plesiosaurs swam by rowing themselves through the water with their paddle-like limbs, and, like the ichthyosaurs, they were undoubtedly fish eaters. They appeared in the Middle Triassic and were common in Jurassic and Cretaceous seas.

Also included in the Sauropterygia are two Triassic groups—the less highly specialized nothosaurs; and the aberrant, armored placodonts with shortened necks and broad, flat teeth for crushing molluscs.

Subclass Synapsida

Except for *Mesosaurus,* the synapsids were the only reptiles to develop lateral but not dorsal temporal openings in the skull. One of the oldest of the reptile groups, they appeared in the Carboniferous and were the commonest reptiles of the Permian. After the emergence of the dinosaurs in the Triassic, the synapsids declined rapidly, and by the close of the period had practically disappeared. But before they did so, they gave rise to the mammals, the group that replaced the dinosaurs to become the dominant animals of the Cenozoic.

The Synapsida are divided into two orders in a more or less horizontal fashion.

Order Pelycosauria. These were the primitive synapsids of the Carboniferous and Lower Permian, with a few stragglers surviving into the Upper Permian and one doubtful form in the Lower Triassic. Most of them were rather small, though some of the later forms were more than 3 m. in length. They all had sprawling legs like the amphibians. Some were long-snouted fish eaters, others were more terrestrial carnivores, and a few became herbivorous.

Most spectacular of the pelycosaurs were the ship lizards (e.g., *Edaphosaurus*), so named because the enormously elongated neural spines are thought to have supported a curious, sail-like fin along the midline of the back. It has been suggested that this large fin may have been a means of

FIG. 5-7. Synapsid reptiles: upper, *Cynognathus;* middle, a pelycosaur, *Dime-trodon;* lower, a dicynodont, *Kannemeyeria.*

controlling body heat. It was probably rich in blood vessels. Spread out in the sun, it would have speeded up the absorption of heat. If the animal grew overheated, it could move into a cooler, shady spot where heat would be quickly dissipated through the membrane. This mechanism if it were truly used for thermostatic control, was not the one that finally proved successful; it remained for another group, with a different sort of thermostatic control, to lead on to the mammals.

Order Therapsida. The order of advanced synapsids appeared in the Middle Permian and its members were enormously abundant and diversified in the Upper Permian and Lower Triassic. Some were carnivorous, others herbivorous. Many therapsids were bulky animals and some attained lengths of more than 4 m. Others were smaller, active, predaceous types. They were characterized, not so much by common structural features as by the trends the various lines showed toward the mammalian condition. These trends included the development of a heterodont dentition, of a double condyle on the skull, and a change in the position of the limbs so that they no longer sprawled sideways but were brought under the body to support it off the ground. Therapsids also developed a secondary palate, a bony plate separating the nasal chamber above from the oral chamber below. Because they do not maintain a constant body temperature and can survive with a low metabolic rate, ectothermic reptiles can tolerate fluctuation in the supply of oxygen, and may depress or even stop their breathing for a time. The endothermic mammals, in which metabolic activity continues at a steady and high rate, must breathe regularly. The development of a secondary palate permits an animal to breathe while it is holding food in its mouth. This device was a major step in evolution toward the mammals.

Since mammalian characters include physiological functions and such soft parts of the body as mammary glands and hair, it is impossible to say from fossil evidence exactly when the therapsids ceased to be reptiles and became mammals. We will leave them on the threshold and return to the modern amphibians and reptiles.

Collateral Reading and General Reference

Colbert, E. H. *The Dinosaur Book.* 2d ed. New York: McGraw-Hill, 1951. (A popular account, recent and readable.)

Piveteau, Jean. *Traité de Paléontologie.* Tome V. *Amphibiens, Reptiles, Oiseaux.* Paris: Masson et Cie., 1955.

Romer, A. S. *Vertebrate Paleontology.* 2d ed. University of Chicago Press, 1945. *Osteology of the Reptiles.* University of Chicago Press, 1956.

REPRODUCTION

AND LIFE HISTORY

OF AMPHIBIA

MOST OF THE ACTIVITIES of an animal, whether it is feeding, seeking a safe resting place, or escaping from an enemy, are directed toward the preservation of the individual. Periodically, however, these activities are superseded by another set leading to the perpetuation of the race. The urge toward reproduction is so strong that the animal may change its customary way of life completely. It may stop feeding, leave its home, and travel long distances, exposing itself to enemies on the way, to participate in the breeding activity of its kind. In this chapter we are concerned with these activities and their outcome, the development of a new generation to replace the old.

The most striking thing about amphibian reproduction is its diversity of pattern. We can hardly make a general statement about any phase of it that does not have some exceptions. Fertilization may be internal or external, the animals may lay eggs or bring forth their young alive, eggs may be laid in water or on land, there may be a larval stage or development may be direct, parents may abandon their eggs or guard them. If we were to speak in anthropomorphic terms, we might say that the amphibians have been experimenting, trying to find the method or methods of reproduction best suited to the new conditions of life on land.

BREEDING ACTIVITIES

Breeding Season

Obviously, before an animal can breed it must be physiologically ready, that is, it must have a supply of ripe ova or sperm ready to be shed. Most animals are in breeding condition for only a relatively short period, or perhaps several such periods, during a year. An innate physiological rhythm apparently governs the development of the amphibian sperm and ova, timed so that breeding will take place when conditions are most favorable for the development of the eggs and larvae.

Seasonal changes in the activity of both the anterior pituitary gland and of the gonads may be involved in the regulation of this rhythm. Males of the Common Frog of Europe (*Rana temporaria*) cannot be induced to produce sperm during the autumn and winter months. Only in the spring do the testes respond to the gonadotrophic hormones of the anterior pituitary by the formation of mature spermatozoa. On the other hand, ovulation can be induced in females of this and many other species of frogs at any time of year by injection of these hormones. That the females normally produce eggs only in the spring and summer indicates that the rhythm is controlled by cyclical activity of the pituitary. It is probable that it is also at least partly under some sort of environmental control. The activity of the anterior pituitary is known to be influenced by such external factors as light and temperature.

Once the animals are in breeding condition, breeding activity is induced by appropriate climatic factors. In the north temperate zone, where most breeding activity takes place in the spring and summer, increase in temperature probably has an important triggering effect. However, breeding is not always associated with a rise in temperature. The Florida race of the Spring Peeper (*Hyla crucifer bartramiana*) spawns during the winter months of December, January, and February, whereas the northern race (*H. c. crucifer*) breeds in April and May. Probably the temperatures at which breeding takes place are very similar for both forms. Most frogs live in the tropics where there is little seasonal change in temperature. In the so-called "wet and dry" tropics, the onset of the rainy season is the most important triggering mechanism in inducing frogs to spawn.

In some species of frogs, the adults in one locality all reach breeding condition at about the same time, in others they do not. If they do, then breeding is explosive, with all the reproductive activity taking place in a few days or weeks. If they do not, then breeding is protracted and may

extend over a period of months. The breeding season for many frogs on
Barro Colorado Island in Panama lasts eight months, the duration of the
rainy season. In warm, constantly moist climates, some species may breed
all year round. Spadefoot Toads (*Scaphiopus*) are explosive breeders. They
lay their eggs in temporary pools produced by torrential downpours; most
of the activity is over twenty-four hours after the onset of such a storm.
These toads cannot be said to have a true breeding season. One year they
may spawn in March, the next year in September; in some years there may
be several breeding periods, in other years none. The triggering mechanism
is unknown; it may be either the sudden decrease in barometric pressure
accompanying these heavy storms or the saturation of the ground with
water. The adults probably remain in a state of physiological readiness for
most of the warmer part of the year.

How many times an individual frog may breed during a season is not
known. *Rana temporaria* in England is an explosive breeder. In a study of
tagged frogs in a single pool during a twelve-day breeding period, one male
was seen clasping three different females on each of three successive nights.
None of the marked females were observed in amplexus more than once. In
forms that breed several times a year, such as *Scaphiopus*, it is not known

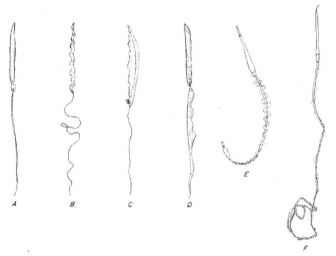

FIG. 6-1. Spermatozoa of some amphibians: A, *Desmognathus phoca;* B,
Cryptobranchus alleganiensis; C, *Amphiuma means;* D, *Triturus marmoratus;*
E, *Salamandra salamandra;* F, *Bombina variegata.* [Redrawn from Noble and
from Angel.]

whether the frogs of the second breeding aggregation are ones that failed to find mates the first time or whether they are the same frogs that bred before. Where the breeding season is protracted, it is possible that a single female may produce more than one clutch of eggs a year, but this is not definitely known.

Salamanders are predominantly animals of the north temperate zone and most of them breed in the spring. But there are many exceptions. In the western United States, the Del Norte Salamander (*Plethodon elongatus*) breeds in the late fall and the California Tiger Salamander (*Ambystoma tigrinum californiense*) in winter.

Breeding Sites

Different species of amphibians show a wide variety in choice of breeding sites. This is particularly true of the frogs. They may lay eggs in open, standing water, either permanent or temporary, in running water (even mountain torrents), on the ground under stones or logs, in burrows, in mud basins constructed by the females, on the undersides of leaves suspended over water, or in water collected at the bases of the leaves of tropical air plants. Frequently the eggs or tadpoles, or both, are carried about by one of the parents. The young of *Rhinoderma darwini* go through their post-hatching development in the vocal pouch of the male!

Salamanders show less diversity in choice of breeding sites, but even so their eggs may be laid in still water, in swift flowing streams, or on land.

Breeding Congregations

A few amphibians, including some aquatic frogs and salamanders and some terrestrial forms that lay their eggs on the ground, normally live during the nonbreeding season in their breeding places. Individuals of other species disperse widely and are found living far from the spawning sites. When the time for reproduction is at hand, they congregate in large numbers in areas much smaller than, and often of a much different character from, the non-breeding habitat. The males of both frogs and salamanders characteristically arrive before the females. These large breeding congregations increase the chances of an individual animal finding a mate.

The question arises of how the widely dispersed animals find their way back to the restricted breeding areas. We cannot assume that a frog re-members the route by which it traveled away from a pond when it emerged from the tadpole state a year or more earlier. Nor can we assume that it is mere aimless wandering that brings thousands of frogs of the same species

FIG. 6-2. The eggs in an opened nest of the terrestrial breeding leptodactylid, *Kyarranus loveridgei*, of Australia. [Photograph by John A. Moore.]

together on the shores of one small lake. The physiological changes that take place when the animal reaches breeding condition apparently sensitize it to certain stimuli to which it does not respond at other times. In view of the multiplicity of breeding sites selected, the stimuli to which the animals respond must differ from species to species. Voice seems to be important in many frogs. The males nearest the breeding site begin calling first. Those farther off are guided by the calls of the earlier arrivals and hasten to join the chorus. A Southern Toad, *Bufo t. terrestris*, will move toward a chorus from a distance of nearly 1 km. Voice is not the only stimulus directing movement in frogs, however. *Rana temporaria* in England has been shown to migrate in response to the odor of ripening algae in the ponds. And voice is certainly not the cue for the silent salamanders, which also migrate over long distances to reach their breeding sites. The odor of marsh vegetation, temperature and humidity gradients, the downhill slope of the ground, and, for aquatic salamanders, oxygen gradients and the flow of current, have all been suggested as stimuli guiding the direction of migration in amphibians.

Voice

Amphibians are the first vertebrates to have true voices. The ability to produce sounds differs in the different groups; caecilians and sirens are essentially silent as are most salamanders and female frogs. Male frogs possess some of the most remarkable voices in the animal kingdom. They have well-developed vocal cords and also large, inflatable vocal pouches to add resonance to their calls. Their habit of congregating in large groups results in choruses that can be heard for great distances and that are almost deafening when one is close by.

Frog calls differ in pitch, duration, frequency, and harmonics. They may be trilled or untrilled, and may consist of a single note or several. Sometimes two or three individuals call in sequence so that a rhythmic structure is imparted to the chorus. Each species has its own particular call which is so characteristic that a frog can be identified by its voice alone just as surely as can a bird (see Fig. 10-4). Furthermore, similarity of voice between different species may be a valuable clue to relationships. Voice is such a striking feature that many frogs are named for the sounds they produce. Thus we have the Bullfrog (said to bellow like a bull), Barking Frog, Bird-voiced Treefrog, Pig Frog, Bell Frog, and Snoring Frog.

The calling of male frogs is closely associated with reproduction. In some species the voices of the earliest males to arrive at the breeding site attract other males as well as females to the same limited area. These are the species that form large choruses. An example is the well-known Spring Peeper (*Hyla crucifer*) of the eastern United States. In other species, only the females are attracted to a calling male, the breeding pairs are scattered over a wider area, and no true chorus results. This is especially true for such terrestrial forms as *Eleutherodactylus*. Sometimes several species resort to the same breeding site at the same time; it is very difficult for man to distinguish individual calls in the resulting cacophony.

Bringing the individuals together in the spawning area is not the only part played by voice during the breeding season. Each male Bullfrog (*Rana catesbeiana*) establishes a calling station, a particular spot to which he returns night after night. His voice apparently warns other males of the same species to stay away from that spot. Here calling by the frogs may perform the same function as the singing of birds in the spring—that of establishing territory. In a mixed chorus where several different species are calling together, voice may serve as an isolating mechanism, preventing the females from going to males of another species. Voice may also be useful in sex recognition, preventing a male from clasping another male rather than a female.

As a rule, only male frogs have well-developed vocal cords but a female may give a mercy scream when caught by a snake and some females will grunt when handled. Occasionally frogs that breed in the spring will also call in the fall. Some, such as the Squirrel Tree Frog (*Hyla squirella*) have distinctive rain calls which differ from the breeding calls and are given away from the spawning grounds in humid weather. For this reason they are often called Rain Frogs. Nonbreeding Spadefoot Toads are sometimes heard calling from underground.

At least one salamander, the Pacific Giant Salamander, *Dicamptodon ensatus,* has true vocal cords. Other salamanders and trachystomes yelp, squeak, or whistle. Many of these noises are produced accidentally but some are not. Their function is not clearly understood. They may be part of a defense mechanism, though in a few species, such as the Tiger Salamander (*Ambystoma tigrinum*) and the European *Salamandra s. taeniata,* there may be some sexual significance.

Sex Recognition

When we speak of sex recognition there is no implication that amphibians consciously distinguish between the sexes. It is simply that the presence of a female evokes from the male a given series of reactions that are not evoked by another male. Similarly a female responds to a male as she does not to another female. We know very little of the precise mechanism of sex recognition. There are, of course, many differences between the sexes beside the primary differences in the genital systems. Some, such as the enlarged tympanum of a male Bullfrog, are permanent, others only appear with the onset of the breeding season. The latter are exemplified by the spiny nuptial pads on the forefeet of many male frogs and toads. These pads help the male grasp and hold the female, but, like the majority of secondary sex characters in amphibians, seem to have little value in sex discrimination.

Both behavior and the odor (or some other recognizable stimulus) from the secretions of hedonic (pleasure-giving) glands probably function in sex recognition by salamanders. During courtship the male may prod, nose, or otherwise rub against another animal, by his actions showing that he is a male. A ripe female accepts these advances, another male resists or runs away. Males of some European newts are brightly colored and display before the females. Sexual difference in color may aid in sex discrimination in these forms.

A female frog responds to the voice of a calling male by moving toward him. Males seem to have little ability to discriminate between the sexes but tend to grasp any object of approximately the right size that approaches. A female whose abdomen is distended with ripe eggs will be held until the eggs are deposited. A male or a female who has already laid will be quickly released. On the other hand, a frog will continue to clasp another male whose abdomen has been artificially distended. A male of the Leopard Frog (*Rana pipiens*) clasped by another male utters a little croak or chirp and is promptly released. The peculiar posture and gait assumed by a female frog during egg deposition apparently help stimulate the male to retain his grasp.

Courtship and Amplexus

Before actual mating takes place, many animals go through a series of actions designed to stimulate one or both partners to perform the sexual act. Procedures of the salamanders vary from a lack of any true courtship activity in the primitive hynobiids to quite elaborate behavior patterns—the liebespiel (love play)—of the more advanced salamandrids and plethodontids.

Males of *Hynobius lichenatus* arrive at breeding ponds first. The females follow a day or so later but are ignored by the males. When a female begins to lay, the male rushes forward and grabs the emerging egg sacs with his forelimbs while pushing the female away with his hind limbs. There is no contact and no mutual stimulation between the pair until egg deposition actually begins.

The male of *Eurycea bislineata*, a plethodontid, noses the female and frequently bends his head across her cheek. She finally responds by straddling his tail and pressing her snout against the glands at its base. The tail is bent sharply aside, and the pair walks along in this fashion until mating is completed. In some species of salamanders the male creeps under the female and carries her along on his back, in others, he lashes his tail to waft the secretions of his hedonic glands toward the female. Because courtship behavior differs from species to species, it probably functions as an.isolating mechanism as well as for sex recognition and stimulation.

Frogs have no liebespiel; the calling of the male seems sufficient to stimulate the female. She indicates her readiness to spawn by approaching and allowing the male to mount her back and clasp her body with his front legs. This is amplexus. He may clasp just behind her front legs (axillary amplexus) or farther back, sometimes just in front of her hind legs (inguinal amplexus). Since he faces the same way she does, he is now in a position to shed sperm over the eggs as they are extruded.

FERTILIZATION AND DEVELOPMENT

Fertilization

The object of the many complex activities discussed above is to bring together the male and female gametes so that they may fuse. This fusion of egg and sperm is fertilization, which may take place externally or internally.

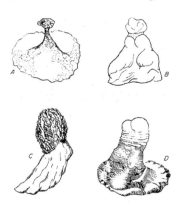

Fig. 6-3.

Salamander spermatophores: A, *Diemictylus v. viridescens;* B, *Ambystoma jeffersonianum;* C, *Eurycea bislineata;* D, *Desmognathus f. fuscus.* [Redrawn from Bishop.]

Fertilization among the primitive salamanders, Hynobiidae and Cryptobranchidae, is external but among all other salamanders it is internal. During courtship the male deposits several little gelatinous packets containing sperm (spermatophores) and the female picks them up with the lips of her cloaca. In the cloaca the gelatinous cap of the spermatophore dissolves and the sperm make their way to the spermatheca, a diverticulum of the roof of the cloaca. Apparently the eggs are usually fertilized as they pass by the spermatheca, but there is evidence that sperm of the European *Salamandra atra* migrate up the oviduct before the eggs are fertilized. On the other hand, the ova of the newt *Diemictylus* cannot be fertilized until jelly layers have been deposited around them by the glands of the oviduct. Sperm high up in the oviduct would be wasted. Most salamanders lay their eggs shortly after mating but *Necturus maculosus*, which mates in the fall, does not deposit the eggs until the following spring.

Frogs usually have external fertilization, with the male shedding his sperm over the eggs as they are deposited by the female in water or in some moist environment. However, the live-bearing frog of Africa, *Nectophrynoides*, practices internal fertilization, although it has no intromittent organ for the transmission of sperm. The "tail" of the American tailed frog, *Ascaphus*, is an extension of the cloaca which serves as an intromittent organ (see Fig. 13-3). Fertilization is internal in these mountain torrent forms, perhaps to prevent the sperm from being swept downstream before the eggs are fertilized.

The male caecilian has an intromittent organ to eject sperm into the cloaca of the female and some caecilians are live bearers. Internal fertilization must be the rule in this group but we know almost nothing of how the male finds the female or where and in what manner mating takes place. Nothing is known about mating among the trachystomes.

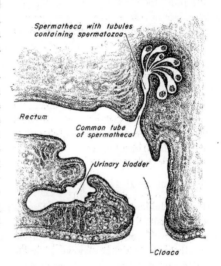

Eggs

Amphibian eggs are generally mesolecithal. But the amount of yolk varies widely within the

FIG. 6-4.

Diagrammatic sketch showing the relation of the spermatheca to the cloaca in a salamander.

groups. Species that lay their eggs on land and undergo direct development have larger yolks than do species that lay in water. Thus the terrestrial egg of the Cliff Frog (*Syrrhophus marnocki*) has a yolk 4 mm. in diameter, whereas the yolk of the aquatic eggs of the Cricket Frog (*Acris gryllus*) is only 1 mm. in diameter, although both adults are about the same size.

There is enormous diversity in the way eggs are deposited by different species of amphibians. Generally, however, glands in the walls of the oviduct secrete a gelatinous substance around them as they pass down to the cloaca. This substance swells in water to form a protective jelly. Sometimes many eggs are enclosed in a single jelly mass, sometimes the eggs are separate.

The jelly may be in the shape of long strings, in one large, irregular packet, or in several smaller packets deposited separately by the female. Eggs may float on the surface of the water, sink to the bottom, be attached to the stems of aquatic vegetation, or to the undersides of leaves, sticks, or rocks in the water. They may also be laid on land.

The numbers of eggs laid varies greatly from species to species. The tiny *Sminthillus*, of Cuba, is said to lay only one egg. On the other hand, some ranas and bufos lay upwards of twenty thousand in one spawning season. The number of eggs depends in part on the size of the animal; there is simply room for more eggs to develop in the body of a female of a large species than in that of a small one. But it also depends on whether the eggs are laid in the open or in some protected spot and whether they are guarded by one of the parents. Species of *Eleutherodactylus*, which lay under rocks and logs on the ground and then guard the eggs, have clutches of less than a hundred. The bufos, which spawn in open water and abandon the eggs as soon as laid, have some of the largest clutches reported for any frog.

In general, salamanders lay fewer eggs than frogs, but they may deposit as many as several hundred. Internal fertilization insures that fewer eggs will be wasted because they are not fertilized, and the salamanders are more likely to guard their clutches.

Developmental History

Most amphibians differ from the other tetrapods in having two definitive developmental periods: embryonic development prior to hatching, and post-hatching development until metamorphosis to the adult form. Animals in the posthatching stage are known as larvae among the salamanders and as tadpoles among the anurans. Occasionally there is no larval period and the young hatch (or are born) as miniature replicas of the adults.

Oviparity, Ovoviviparity, and Viviparity. Usually the young of amphibians hatch after the eggs have been laid and the animals are said to be oviparous. Rarely, however, the eggs are retained in the body of the female while they pass through their embryonic development and the young are "born alive." If the developing embryo in the mother's body is nourished entirely by food stored in the yolk of the egg, the animal is ovoviviparous. If the embryo obtains part of its food from maternal tissues (the walls of the Müllerian duct) the animal is viviparous. The European *Salamandra* is ovoviviparous and bears living larvae. The larvae of one species (*S. atra*) metamorphose before birth. The African frog *Nectophrynoides* is also ovoviviparous and the young metamorphose before they are born. Some caecilians exhibit a specialized form of viviparity. The young are nourished by oil glands in the lining of the oviduct and, like those of *Nectophrynoides*, resemble the adults when born. Other caecilians are oviparous.

Embryonic Development. The early development of the amphibians is typical for vertebrates that have mesolecithal eggs. The stages are: early cleavage, blastula formation, gastrulation, neurulation, tail-bud stage, the period of organogeny (formation of organs), the period of early heart beat and development of gill buds, and finally the development of gill circulation and hatching. The details of these stages can be found in any text on embryology and need not be discussed here. Figure 6-5 shows the external morphology of a developing salamander, *Necturus;* stage J is about comparable to hatching. Table 6-1 lists comparable stages in the development of

TABLE 6-1.   DEVELOPMENT STAGES OF *Rana pipiens* (From Rugh, *The Frog*, 1951.)

Stage	At 18°C	At 25°C
Fertilization	0 hours	0 hours
Gray crescent	1 hour	$\frac{1}{2}$–1 hour
Rotation	$1\frac{1}{2}$ hours	1 hour
Two cells	$3\frac{1}{2}$ hours	$2\frac{1}{2}$ hours
Four cells	$4\frac{1}{2}$ hours	$3\frac{1}{2}$ hours
Eight cells	$5\frac{1}{2}$ hours	$4\frac{1}{2}$ hours
Blastula	18 hours	12 hours
Gastrula	34 hours	20 hours
Yolk plug	42 hours	32 hours
Neural plate	50 hours	40 hours
Neural folds	62 hours	48 hours
Ciliary movement	67 hours	52 hours
Neural tube	72 hours	56 hours
Tail bud	84 hours	66 hours
Muscular movement	96 hours	76 hours
Heart beat	5 days	4 days
Gill circulation	6 days	5 days
Tail fin circulation	8 days	$6\frac{1}{2}$ days
Development of operculum	9 days	$7\frac{1}{2}$ days
Operculum complete	12 days	10 days
Metamorphosis	3 months	$2\frac{1}{2}$ months

the Leopard Frog, *Rana pipiens*. As the table shows, there is some variation in developmental rate, determined, at least in part, by temperature. Hatching takes place at about the time the gill circulation develops, or at about 6 days at 18°C (Fig. 6-6).

Posthatching Development. The degree of development reached at hatching varies greatly in the amphibians. Some frogs and salamanders pass

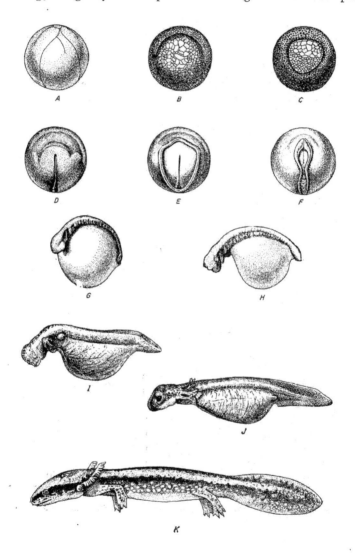

through metamorphosis in the egg or hatch in an advanced condition so that they are nearly ready to metamorphose. Most frogs hatch out as tadpoles, animals quite different in appearance from their parents, and must go through a continuing period of development before they are ready to metamorphose.

The early tadpole of *Rana pipiens* is a small, poorly developed animal, seemingly ill-equipped to meet the vicissitudes of life outside of the protecting egg. Its mouth has not yet broken through, its eyes are not formed, its gills not fully developed. For a few days it remains attached to some object by a pair of V-shaped suckers below the indentation where its mouth will be and draws nourishment from its remaining supply of yolk. A small, round mouth armed with horny teeth and jaws appears, the eyes take shape, the gills grow rapidly, and the tail elongates. The tadpole is soon ready to take up a free-swimming, food-finding existence, but it still does not look much like a frog. An operculum grows back to cover the external gills, which are replaced by a new set of gills. Gradually the legs develop. The hind legs push out as tiny buds, and later develop toes and joints. The front legs are growing at the same time but are hidden by the operculum. Shortly before metamorphosis the tadpole comes frequently to the surface to gulp air. Its lungs are preparing to take over from the gills the task of supplying oxygen to the blood. The long coiled intestine necessitated by the vegetarian diet of the tadpole shortens, for the adult frog is a flesh eater. The front legs push out, the left through the spiracle, the right through the opercular wall. The gills degenerate, the horny jaws and teeth are shed along with the tadpole skin, the round mouth widens, the tail is resorbed, although a vestige still remains, and the little animal, now a frog, hops out on land.

FIG. 6-5. The development of *Necturus maculosus*. A. Side view of egg 1 day and 8 hours after deposition, showing second and third cleavage grooves. B. Bottom view of egg 6 days and 16 hours old. The crescentic blastopore lip sharply separates the large yolk cells from the small cells of the blastodisc. C. Bottom view of egg 10 days and 10 hours old, showing large circular blastopore. D. Top view of egg 14 days and 4 hours old. Blastopore smaller. The beginning of neural fold formation, especially anteriorly. E. Top view of egg 15 days and 15 hours old. Yolk plug still visible. Neural fold prominent. Its free ends reach nearly to the blastopore. F. Top view of egg 18 days and 15 hours old with three or four pairs of body segments visible. G. Dorsolateral view of embryo 22 days and 17 hours old; length 8 mm.; 16 to 18 body segments. H. Side view of embryo 26 days old; length 11 mm.; 26 to 27 body segments; eye, ear, nasal pits, and mouth well defined. I. Side view of embryo 36 days and 16 hours old; length 16 mm.; 36 to 38 body segments. J. Side view of larva 49 days old; length 21 mm. K. Side view of larva 97 days old; length 34 mm. [From *The Biology of the Amphibia*, by Gladwyn K. Noble; reprinted through permission by Dover Publications, Inc., New York 14, New York ($2.98).]

100

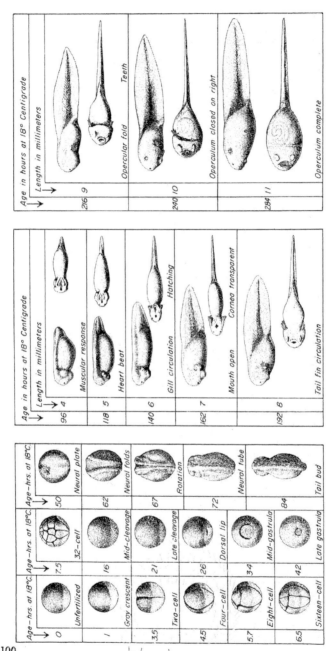

FIG. 6-6. The external morphological features of the developing embryo of the frog, *Rana pipiens*. [Redrawn from Shumway.]

Salamander larvae are quite different from tadpoles. The frog tadpole has one or a pair of suckers on the ventral side of the head from before to shortly after hatching; these are lacking in salamander larvae. On the other hand, the larvae of many salamander species have balancers, long, rod-like structures, one on each side, that protrude from the sides of the head below the eyes. The salamander larva has three pairs of gills, the frog tadpole only two well-developed pairs, the upper or third pair being rudimentary. After hatching the tadpole develops an operculum covering the gills; gills are usually plainly evident in salamander larvae until the time of metamorphosis. The legs of salamander larvae are often well developed at hatching. The

SALAMANDER LARVA   FROG TADPOLE

1. SHORTLY BEFORE HATCHING

Long body
No adhesive organs
Three pairs of gill rudiments

Shorter body
Prominent adhesive organs
Only two pairs of gill rudiments distinct

2. SHORTLY AFTER HATCHING

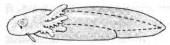

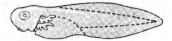

Long body
No adhesive organs
Balancers present in some species
Mouth becomes large
Three pairs of well developed external gills,
  the uppermost typically the longest
Opercular folds simple

Short body
Prominent adhesive organs
No balancers
Mouth remains small
Only two pairs of well developed external gills,
  third pair (uppermost) rudimentary
Opercular folds develop rapidly, soon cover gills

3. LARVAL CHARACTERS FULLY DEVELOPED

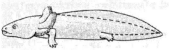

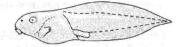

Long body
Mouth large, with simple labial folds
External gills present
Opercular folds open and loosely overhang
  sides of neck; no spiracle
Foreleg buds external, clearly visible
  throughout their development

Short, egg-shaped head and body unit
Mouth small, with complex accessory parts
No external gills
Operculum closed except for a small spiracle
  draining gill chamber
Foreleg buds internal, concealed under closed
  operculum until metamorphosis

FIG. 6-7. Comparison of salamander larvae and frog tadpoles. [After Orton.]

mouth is like that of the adult. The mature salamander larva resembles the adult much more closely in body proportions than the tadpole resembles the frog.

Larval Types. Differences in posthatching habitat are correlated with morphological differences in the developing larvae. Salamanders have three basic types of life histories: some are specialized toward breeding in open bodies of quiet water; some are modified for a mountain stream existence; some have become specialized for terrestrial life.

Warm, still pond water has relatively little oxygen dissolved in it; therefore a pond-type larva needs large respiratory surfaces to absorb enough oxygen for its bodily needs. For example, the pond-type larva of *Ambystoma* has long, very filamentous gills, and a well-developed tail fin extending up over its back to form a dorsal body fin.

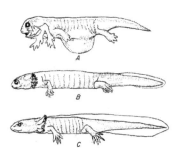

In the cool, rapidly flowing water of mountain torrents, plenty of oxygen is available. The problem faced by larvae in this environment is rather to keep from being swept downstream by the current. For this reason the brook-type larva of *Dicamptodon ensatus* is streamlined, with a muscular tail, no dorsal body fin, a reduced tail fin, and small gills.

FIG. 6-8.
The principal types of salamander larvae: A, terrestrial type, *Plethodon vandykei;* B, mountain brook type, *Dicamptodon ensatus;* C, pond type, *Ambystoma gracile.* [From *The Biology of the Amphibia*, by Gladwyn K. Noble; reprinted through permission by Dover Publications, Inc., New York 14, New York ($2.98).]

Finally, the terrestrial salamanders, such as *Plethodon,* have a tiny, short-bodied larva with gills developed but lacking filaments. These gills reduce before hatching. The larva has a rather well-developed yolk mass in the gut and a short tail with no fin. These tiny forms hatch out of eggs laid on land and soon become entirely terrestrial salamanders without ever going through an aquatic larval stage.

Because tadpoles are found in a wider variety of habitats than salamander larvae, they also show more morphological diversity. Here again we find a pond type, the typical, deep-bodied, high-finned pollywog. Many mountain brook tadpoles are streamlined, with strong tails and reduced fins. The tadpole of the genus *Staurois* from the mountains of southeastern Asia has an adhesive disc on the ventral surface by which it clings to rocks and so avoids being swept away by the current. Other mountain brook tadpoles have large suctorial lips. Some torrent tadpoles have flattened bodies adapted for gliding over wet stones.

Some species of tree frogs (*Hyla*) lay their eggs in the tiny pools of water held in the axils of the leaves of bromeliads. The tadpoles are slim, with long, whiplike tails and reduced gills. Apparently they get most of their oxygen directly from the atmosphere. They feed on frog eggs, either of their own or other species, and probably also on other tadpoles. Their jaws are strong, but the larval tooth rows have been reduced. Tadpoles of the African *Hoplophryne* develop in similar habitats and show similar modifications.

Embryos of frogs that lack a free-living larval stage may yet show most of the typical tadpole structures, such as the operculum and the larval teeth and jaws. In others these structures are reduced and tend to disappear. The developing young of *Rhinoderma darwini* are distinctly tadpole-like, with closed operculum, spiracle, coiled intestine, and larval mouth parts, although the teeth and jaws never harden.

Duration of Larval Stages. The length of the larval period varies widely, not only from species to species, but also within a given species. As with the eggs, higher temperatures accelerate the rate of development. Factors other than temperature are known to affect developmental rate of *Scaphiopus*. The eggs are laid in temporary pools and development is very rapid since the toads must metamorphose before the pools dry up. When some tadpoles from a drying pond were transferred to an aquarium containing pond water at the same temperature, they remained tadpoles for nearly a month after those from the pond had emerged. Overcrowding, a heavy concentration of chemicals in the water, an increase in the available food supply, probably all speed up the time of metamorphosis.

Spadefoot tadpoles have been known to metamorphose as little as 12 days after egg deposition. Most amphibians probably remain from a minimum of a few months to a maximum of a year in the embryonic and larval stages. However, the Bullfrog (*Rana catesbeiana*) usually spends 14 to 16 months as a tadpole and may not metamorphose until after its third winter. The Mudpuppy (*Necturus*) has been reported to remain four or five years in the larval stage.

Paedogenesis and Neoteny

Trachystomes, proteid salamanders, and some plethodontids never complete metamorphosis. They retain their gills and other larval structures even after they become sexually mature. This is paedogenesis, a genetically fixed condition in which the tissues fail to respond to the secretions of the thyroid gland that bring about metamorphosis in other forms.

Some salamanders, including many ambystomatids, fail to metamorphose when environmental conditions, such as low temperatures or lack of iodine in the water, inhibit the action of the thyroid. They become sexually mature and reproduce in the larval state. In contrast to the paedogenic forms, they are still capable of metamorphosing if the environmental conditions are changed. When such animals are fed thyroid extract, or are simply trans-ferred to water with a higher iodine content, they soon metamorphose. Failure to metamorphose because of environmental conditions is called neoteny.

GROWTH AND LONGEVITY

After metamorphosis, amphibians must still go through a period of growth and development before they become sexually mature. Some species reach

TABLE 6-2.   LONGEVITY OF SOME SPECIES OF AMPHIBIANS

Caudata	Maximum age (years)
Cryptobranchidae (*Megalobatrachus japonicus*)	55
Ambystomatidae (*Ambystoma maculatum*)	25
Salamandridae (*Triturus palustris*)	28
Amphiumidae (*Amphiuma means*)	27
Proteidae (*Proteus anguinus*)	15
Trachystomata	
Sirenidae (*Siren lacertina*)	25
Anura	
Pipidae (*Xenopus laevis*)	15
Discoglossidae (*Bombina bombina*)	20
Pelobatidae (*Pelobates fuscus*)	11
Leptodactylidae (*Leptodactylus pentadactylus*)	12
Bufonidae (*Bufo terrestris*)	31
Hylidae (*Hyla caerulea*)	16
Ranidae (*Rana catesbeiana*)	16
Rhacophoridae (*Kassina weali*)	9
Microhylidae (*Kaloula pulchra* and *Microhyla carolinensis*)	6

breeding age in about a year. Others, particularly the larger forms like the Bullfrog (*Rana catesbeiana*) and the Mudpuppy (*Necturus maculosus*), are not ready to breed until two or three years after metamorphosis.

Growth does not stop when an amphibian reaches sexual maturity. Many, perhaps all, continue to grow, although very slowly, for the rest of their lives.

Maximum age is very difficult to determine for animals living under natural conditions. It is much easier to obtain accurate figures on animals reared in captivity. But such sheltered forms probably survive much longer than wild animals. The figures in Table 6-2 are for captive specimens and do not indicate longevity under natural conditions. Although the table is incomplete it does indicate that some of these small animals are capable of living for surprisingly long times under favorable conditions. It also shows that the sluggish salamanders apparently live longer than do the more lively frogs.

We know nothing of the reproductive habits of the early amphibians but it is doubtful if all the diversity shown by the present-day forms is of recent origin. The ancestral groups must at least have had considerable genetic plasticity to have allowed such diversity to develop. Most of the early forms failed to find a really satisfactory solution to the problem of reproduction on land and their descendants have remained amphibians. But members of one group, at least, did evolve a practical method of terrestrial reproduction by means of the amniote egg, and they became the reptiles.

Collateral Reading and General Reference

Angel, F. *Vie et Moeurs des Amphibiens*. Paris: Payot, 1947. (A semipopular account of the habits of amphibians, cosmopolitan in scope.)

Cochran, D. M. *Living Amphibians of the World*. New York: Hanover House, 1961. (A beautifully and profusely illustrated account of the amphibians, with much information on life histories.)

Mertens, R. *The World of Amphibians and Reptiles*. Translated by H. W. Parker. New York: McGraw-Hill, 1960. (A fascinating, splendidly illustrated book, written by one world-famous herpetologist and translated by another.)

Noble, G. K. *Biology of the Amphibia*. New York: McGraw-Hill, 1931. Reprinted by Dover Publications, 1954.

Oliver, J. A. *North American Amphibians and Reptiles*. Princeton, New Jersey: Von Nostrand, 1955. (A good, recent natural history of the herptiles of the United States and Canada.)

Rugh, R. *The Frog: Its Reproduction and Development.* Philadelphia: Blakiston, 1951.

Smith, M. *The British Amphibians and Reptiles.* Rev. ed. London: Collins, 1954. (The best natural history accounts available, though limited to the fauna of the British Isles.)

REPRODUCTION

AND LIFE HISTORY

OF REPTILIA

WHEN A GROUP of animals makes a major shift from one mode of existence to another, there must be a close correlation in evolutionary changes of structure, function, and activity. Nowhere is this more marvelously shown than in the reproductive pattern that allowed the reptiles to free themselves completely from the need to return to the water to breed. Terrestrial breeding habits, internal fertilization, shell, and amnion go hand in hand to produce the truly terrestrial egg of the reptile.

Sperm can travel only in a fluid medium. When fertilization takes place away from water, the fluid carrying the sperm must be protected from desiccation. The best method for insuring this is emission of the seminal fluid directly into the reproductive tract of the female by the male —the act of copulation. Since the ovum is fertilized before it leaves the body of the female, it can be provided with a protective shell deposited by the glands of the oviduct.

The ovum also needs a fluid medium in which to develop. The amnion encloses the embryo in a fluid-filled sac. Caecilians, most salamanders, and a few frogs adopted internal fertilization, but they never developed a shelled, amniote egg to go with it and so have been unable to exploit to the full the possibilities of this more efficient method of fertilization.

DIFFICULTIES IN THE STUDY OF
REPTILIAN REPRODUCTION

Reproduction of reptiles is more difficult to study than is that of most amphibians. Reptiles do not, as a rule, come together in large breeding aggregations, nor do they advertise their intentions as do the vociferous frogs. It is thus far more difficult for the herpetologist to be in the right place at the right time to observe reptilian mating. Information gathered from zoo specimens is helpful but is always questionable because the conditions of captivity may have modified the behavior of the animals.

Two other phenomena complicate the study of reproduction in the reptiles. These are delayed fertilization and the initiation of development before egg deposition.

Delayed Fertilization

Among most vertebrates, fertilization takes place shortly after the emission of the sperm by the male, and indeed the sperm lose their ability to activate the ova and initiate development after a few hours or a few days at most. In many reptiles (and a few salamanders) this is not so. Sperm may be stored in an inactive state in the female reproductive tract and may retain their ability to fertilize the ova for months or even years. Delayed fertilization is sometimes called amphigonia retardata.

Table 7-1 shows the length of time ten female Diamondback Terrapins (*Malaclemys*), had been separated from males, the number of eggs laid

TABLE 7-1.   VIABILITY OF SPERM OF *Malaclemys* AFTER SEPARATION OF MALES AND FEMALES

Length of separation (years)	Number of eggs laid	Number fertile
1	124	123
2	116	102
3	130	39
4	108	4

each year, and the number of fertile eggs. There was a decided decline in the number of fertile eggs after two years, but a few of the sperm were viable at four years. Female Box Turtles (*Terrapene carolina*) have also laid fertile eggs two, three, and four years after separation from males.

Delayed fertilization is apparently rare in lizards since it has been reported only in the chameleons. An African Dwarf Chameleon (*Microsaura p. pumila*) produced five eggs, all fertile, five months after the last possible copulation.

Sperm have been found to remain viable for various lengths of time in the female reproductive tract of several snakes. Table 7-2 gives some examples.

TABLE 7-2.   VIABILITY OF SPERM IN FEMALE SNAKES AFTER SEPARATION FROM MALES

Species	Duration of viability
Causas rhombeatus	5 months
Drymarchon corais	4 years, 4 months
Leptodeira annulata polysticta	5 years
Storeria dekayi	3 months
Thamnophis s. sirtalis	3 months

Histological studies of the oviducts of female rattlesnakes (*Crotalus*) show that, from posterior to anterior, the reproductive canal is divided into an expanded vaginal portion, a tightly coiled smaller portion, a rather glandular uterine portion, a tightly coiled tube, and finally the ostium nearest to the ovary. Microscopic examinations show that stored sperm are consistently located in the coiled portion of the oviduct between the vaginal

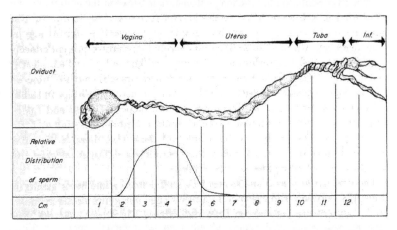

FIG. 7-1. Diagram illustrating the relative concentration of sperm in the various regions of the "ripe" female oviduct of the Prairie Rattlesnake, *Crotalus v. viridis*. [After Ludwig and Rahn.]

expansion and the glandular uterine portion. The sperm do not penetrate beyond the compound glands marking the beginning of the uterus proper. It may be that secretions from these glands inhibit passage of the sperm into the anterior portion of the oviduct until just before ovulation. The sperm must then migrate into the anterior oviduct in order to fertilize the eggs.

Oviparity, Ovoviviparity, and Viviparity

Most reptiles are oviparous. Some lizards and snakes are ovoviviparous, with the eggs hatching either in the oviduct or just after they are laid. There is really no sharp distinction between oviparity and ovoviviparity, since many oviparous forms lay eggs in which the embryos have already started to develop. Eggs of some colubrid snakes, examined at the time of deposition, contained embryos ranging from 15 to 55 mm. in length. Such eggs may not hatch for two or three months. Alligator eggs also seem to go through some development in the oviduct before deposition. This is also true for some of the spiny lizards (Sceloporus).

Of two species in a single genus, one may be oviparous, the other ovoviviparous. Snakes of the genus Natrix lay eggs in Europe and bear living young in North America. The condition may not even be fixed in a single species. The European lizard, Lacerta vivipara, which bears living young throughout most of its range, lays eggs in the Pyrenees.

True viviparity implies exchange of materials between the embryonic and maternal bloodstreams. This condition is approached by a few of the Squamata. Lizards of the family Xantusidae form a shell around the egg in the oviduct but it soon disintegrates and a kind of primitive placenta develops. This actually provides nourishment for the embryo which nearly doubles in weight between the time the shell disappears and the time of birth. Similar placenta formation has been observed in the European skink, Chalcides ocellatus, in Australian skinks of the genera Tiliqua and Lygosoma, and also in the Australian snake, Denisonia. Histological studies of the uterine portion of the oviduct of Denisonia show that there is a partial degeneration of the maternal epithelium, permitting diffusion between the bloodstreams of the mother and the embryo.

The oviducts of lizards and snakes lack well-defined albuminous glands in the upper portions. The albumen in a bird's egg contributes to the amniotic fluid that protects the embryo from desiccation. Terrestrial eggs not provided with a sufficient supply of albumen must be laid in surroundings moist enough to allow water to enter through the permeable shell. Such eggs swell noticeably after deposition. (Even eggs that have albumen are permeable and may lose too much water and desiccate in an arid environment.) It is fascinating to speculate that the tendency of lizards and snakes to retain the eggs in the constantly moist environment of the oviduct for varying lengths of time, leading to ovoviviparity and even viviparity, may be a compensatory mechanism for the lack of albumen. Such a mechanism may have been a step in the development of the mammals from the reptiles (though not, of course, from the Squamata).

Amplexus, egg deposition, fertilization, and the beginning of embryonic development follow one another so closely in most frogs that a single series of observations may give us a fairly clear outline of the reproductive pattern of a species. The pattern of reptiles is not as clear-cut. Copulation may take place in the fall, but egg deposition may not occur until the following spring. Even when eggs are laid shortly after mating, it is still possible that the ova were fertilized by sperm received in a previous copulation. The initiation of development does not necessarily coincide with the time either of copulation or of egg deposition so that it is not possible to calculate from one or the other the length of the embryonic period. Long, patient hours in the field, the dissection of many specimens in the laboratory, and much luck are necessary for the study of the life history of a reptile.

BREEDING HABITS

Sexual Encounters

One of the advantages of terrestrial reproduction over the aquatic reproduction of most amphibians is that the egg clutches can be more widely dispersed. When all the females of a population deposit their eggs in a few places, a single catastrophe, such as the too-rapid drying of some temporary ponds, can bring to naught the reproductive activity of a whole season. Heavy concentrations of developing young bring heavy concentrations of predators. And competition between the young for the available food may be very intense. Reptiles do not need to lay as many eggs as frogs to insure survival of enough young to maintain the population.

But terrestrial reproduction does pose another problem. Except for the sea turtles, most reptiles do not come together in restricted areas to form large breeding aggregations. A male must find one or more receptive females in a population that may be widely dispersed. Because of this, the advantages of amphigonia retardata are clear. If a female *Malaclemys* has mated one year, her ability to lay fertile eggs the following year does not depend on her being found again by a male. Moreover, copulation need not take place only at the time of ovulation. Some species of lizards and snakes that normally mate in the spring, may also copulate in the fall, with·deposition of fertile eggs the following spring. Two captive females of the European Ringed Snake (*Natrix natrix helvetica*), normally a spring breeder, mated in September and laid fertile eggs the following March.

Many northern snakes hibernate in communal dens. One such den, in an ant hill, contained 148 Smooth Green Snakes (*Opheodrys vernalis*), 101 Red-bellied Snakes (*Storeria occipitomaculata*), and 8 Great Plains Garter Snakes (*Thamnophis radix*). When snakes emerge from hibernation in the spring, they tend to remain for awhile in the vicinity of the den. Numbers of both sexes are thus present in the same restricted area, it is easy for males to find females, and most mating takes place at that time. In warm climates where snakes do not hibernate, males probably find the females simply by wandering.

Many lizards live in loosely knit colonies where the chances of the two sexes encountering each other are good. The circumscribed habitat of the freshwater turtles also increases the chances of the male finding a female. Males of the terrestrial tortoises seem to wander more during the breeding season than at other times.

Female marine turtles gather in large numbers at certain beaches to lay their eggs. The males congregate just off shore and mate with the females as they leave the beaches. Thus the sperm to fertilize next year's eggs are received at the time the current year's eggs are laid.

Sexual Dimorphism

The degree of sexual dimorphism varies in different orders of reptiles. Among the turtles, the plastron of a sexually mature male may be more hollowed out behind than that of the female. This undoubtedly helps him retain his balance on the convex carapace of his mate during copulation. Tails may also vary in length, those of males usually being longer. The male of some species has long claws on his forefeet with which he tickles the cheeks of the female during courtship. Females are often much larger than males, but males are occasionally larger.

Morphological differences between male and female lizards are often quite apparent. The male is apt to be larger and may have better developed dorsal or caudal crests. Males of some species display brilliantly colored throat fans during the breeding season. An example is the common Anole (*Anolis carolinensis*) of the southeastern United States; a bright green male spreading a bright red throat fan is a spectacular sight. Sometimes there are differences in body color. Males of *Sceloporus undulatus* have more brilliant blue patches on the sides than do females. Male lizards may also have femoral or preanal pores and modified scales in the region of the vent. The true chameleon (*Chamaeleo*) has horns or dermal protuberances near the end of the snout or in front of the eyes which are better developed in males than in females.

Sexual dimorphism is usually much less evident in snakes. Sometimes no distinguishing characters are visible. The main differences are in the number and form of the scales on the ventral side of the body and in the length of the tail. Males of the European Common Viper (*Vipera aspis*) have from 134 to 158 abdominal scales and from 32 to 49 subcaudals (scales on the underside of the tail). Females have 141 to 169 abdominals and 30 to 42 subcaudals. The body of the male is thus relatively shorter and the tail relatively longer than those of the female. The base of the tail in the male is somewhat more dilated to accommodate the hemipenes. The scales of the male are often more heavily keeled. All the scales of the male *Homalopsis buccata*, an aquatic Asiatic snake, are much more strongly keeled than are those of the female. Males of *Chironius carinatus* of Central and South America have the median rows of vertebral scales strongly keeled whereas those of the female are only slightly, if at all, keeled. Among the Boidae, the vestigial hind limbs, or spurs, are better developed in the males than in the females, which sometimes lack them entirely. Tubercles formed of dermal papillae are scattered over the chin and sometimes the sides of the face of the males of a few snakes, such as *Natrix rhombifera* of North America and *Aspidura trachyprocta* of Ceylon. Occasionally one sex or the other is more brightly marked, especially among the vipers. There is no general rule for determining the sex of any snake by external examination. Each species must be studied separately.

Sex Recognition

In general, the female reptile plays a relatively passive role in sexual encounters. It is the male that actively pursues and courts, and hence must be able to distinguish between the sexes. There is evidence that for many reptiles the male distinguishes between "other male" and "not male" rather than recognizing a female as such. That is, another animal, or even an inanimate object, not recognized as a male, is regarded as female. The nocturnal Banded Gecko (*Coleonyx variegatus*) depends primarily on behavior as a means of sex recognition. Anesthetized males were treated as females by other males. When the blue patches on the sides of males of the diurnal Fence Lizard (*Sceloporus undulatus*) were painted out, they were treated as females by males. Males of marine turtles are sometimes lured into the nets of fishermen by wooden decoys shaped roughly like females.

Appearance, behavior, and odor all play a part in the sex recognition of the reptiles. Head bobbing, a characteristic gesture that has been observed in many lizards, snakes, and turtles, seems to function in sex recognition. Chemical stimuli are important among snakes. A male snake will follow the trail left by drawing the skin of a female across a plate but will not follow one made by smears from her cloacal glands. Among diurnal lizards with marked sexual dimorphism, sight probably plays the major role but odor may also be important. We have watched a male Broad-headed Skink (*Eumeces laticeps*) trail a female across the ground, though she was hidden from sight in a hollow tree.

Some male lizards establish a territory which they defend against other males. During the breeding season a male Anole spreads his throat fan and bobs up and down at sight of an intruder. If the latter is another male, it either retreats or displays its fan in return. If both continue to display, a fight may ensue; the loser is driven from the area. A female has no throat fan to display and her failure to respond as a male apparently results in her being recognized as a female and either ignored or courted.

The roaring of a bull alligator in the spring, a characteristic sound of the southern marshes now too seldom heard, may serve both to warn away other males and to attract females.

The combat dance of male snakes may also play a part in sex recognition. Prior gives a vivid account of the dance of a pair of vipers (*Vipera berus*).

"In May I was witness to a peculiar dance, the participants of which were two male adders, which took place in a shallow leaf-carpeted ditch. I was drawn to the spot by the rustling of dead leaves and saw the silver-grey, black-blotched bodies of two male adders writhing and coiling about each other in a fantastic 'dance'. Both would sway sideways in contrary directions, then slowly sweep back, the bodies meeting and then crossing so that each ultimately reached the position occupied by the other.

"After a series of these slow movements, the contestants became suddenly excited, darting and ducking until one was in a position to force his opponent to the ground. While these rapid head movements were being executed the snakes' bodies were coiled in a succession of undulating patterns. After regaining his breath the fallen viper would erect himself again, sidle up to his conqueror and incite him to further strife. I watched the monotonously reiterated posturings until the smaller of the two snakes freed himself and made off with amazing celerity, closely pursued by the other until out of sight. The female, the cause of this, was lying coiled close at hand."

Similar dances during the mating season have been observed in a number

FIG. 7-2. The combat dance of the Eastern Cottonmouth, *Agkistrodon ·p. piscivorous,* in a Florida marsh. [From Carr and Carr.]

of other species. Some keep their bodies in a horizontal position, others, as these vipers, raise the fore part of the body vertically. At first it was assumed that the dancers were a courting pair, but every time such couples have been collected, both individuals have been found to be males. Just what the real significance of the dance is, whether it is an attempt to establish territory, or combat for a female, has not been definitely determined, but it does seem to have some sexual significance.

Courtship

Many reptiles have a more or less elaborate liebespiel that usually precedes actual copulation. Male turtles of the genera *Chrysemys* and *Pseudemys* court the female in the water. Swimming backwards in front of her, or swimming above her, the male vibrates the long claws of his front feet around her cheeks. This is called titillating. Other turtles may bob their heads and nudge and bite at the female.

Among diurnal lizards, particularly those in which the males are brightly marked, the male may posture and display in front of the female. Lizards also indulge in nipping and nudging. Sometimes a pair will walk along,

FIG. 7-3. Courtship in the King Cobra, *Ophiophagus hannah,* showing how the male and female glide along side by side. [Photograph courtesy New York Zoological Society.]

the male straddling the tail of the female and resting his head on her pelvic region when she pauses.

Male and female snakes may glide along side by side, the male caressing the female with his chin and flickering his tongue over her body. He may twine his tail around hers and lie on top of her while his body undulates in a series of waves passing from his tail toward his head (caudocephalic waves).

Unlike frogs, most reptiles do not have voices and in only a few of them does voice play a part in courtship. Many snakes hiss, and one genus, *Pituophis,* has a special membrane on the epiglottis which vibrates when air is expelled from the lungs to produce a peculiar and characteristic fluttering sound. Geckos have well-developed voices; some can be heard for distances of nearly 100 m. But all of these sounds are apparently warning or threatening, not amatory. The voice of the crocodilian, on the other hand, is definitely associated with breeding activity, though whether it is primarily a love call to the female, or a challenge to other males, or both, is not clear. A fundamental note of 57 vibrations per second typically evokes a roar of response from a half-grown American Alligator. Indeed, alligators in a lake near the University of Florida are often stimulated to call by fireworks set off on the campus. The primitive New Zealand Tuatara (*Sphenodon punc-*

tatus) has been heard croaking on cold, misty nights, but again the reason for the call is unknown. It is noteworthy that the best developed reptilian voices are found among nocturnal forms in which hearing is presumably better developed than sight.

Some turtles also have voices and among the testudines, at least, voice seems to be associated with reproductive activity. The Galapagos tortoise bellows while mating. The male of the South American *Testudo denticulata* makes a noise while pursuing the female and while copulating that, to anthropomorphic ears, sounds suspiciously like a chuckle.

Copulation

Courtship is usually, but not always, followed by copulation. Sometimes copulation takes place with little or no preliminary liebespiel, the male simply grasping the female as soon as he sees her.

Aquatic turtles mate in the water, the others on land. The male approaches the female from behind and mounts her carapace, gripping the front part of her shell with his forefeet, and sometimes biting her head and neck. The hind part of his body is pushed down below her carapace. The position looks very awkward, particularly for a species with a high, domed shell, and sometimes the male loses his balance and topples off.

The positions taken during copulation by crocodilians, lizards, and snakes are much alike. The male lies beside or above the female and thrusts his tail under hers so that the regions of the vents are brought together. Among crocodilians, lizards, and some snakes, the male grasps the head, neck, or shoulder of the female with his jaws. Since a receptive female does not struggle to escape, this probably simply helps the male keep his place. Snakes and lizards evert the hemipenis on the side next to the female and insert it in her cloaca. It becomes turgid and, held in place by its spines and calyces, usually cannot be removed forcibly without damage to the animals. Copulation may be completed in a few minutes or it may last for several hours or even a day. Couples in captivity have been known to copulate at repeated intervals over a space of several days.

EGGS AND EMBRYOS

Egg Deposition

Some time after copulation, usually several weeks to several months, the female deposits her eggs. Some reptile eggs are nearly round, others are ovoid. The shells are either leathery or hard and calcareous. Eggs must be laid on land, in places with optimum moisture and sufficient heat for incubation. In general, reptiles are more careful in selecting or constructing suitable sites for eggs than most of the amphibians. We see in them an early stage of the development of the elaborate nest-building–incubation–parental-care behavior that characterizes the birds.

The female Musk Turtle (*Sternothaerus odoratus*) may scatter her eggs about in shallow excavations or leave them only partly hidden in leaf mold or vegetable debris. But most turtles dig holes in which to bury their eggs. The female scoops out an excavation with her hind feet, deposits the eggs, then scrapes the dirt back into the hole. She may pound it down with her plastron and carefully scuff over the surface so that all signs of digging are obliterated.

Snakes and lizards usually lay their eggs in or beneath rotten logs, in hollows under rocks, in burrows, or in holes dug in the earth. Sometimes several females will select the same place so that many eggs will be found together. In Wales, forty bundles of about thirty eggs each were once found in a hole in an old wall. These were eggs of the Ringed Snake (*Natrix natrix helvetica*).

The female American Alligator actually builds a nest for her eggs. She scrapes together vegetable trash into a large pile which may be 2 m. in diameter and nearly 1 m. high. She then scoops out a hole in the top, lays her eggs, covers them over with some of the material, and packs it down. Other crocodilians may either build nests or heap up piles of sand in which to deposit their eggs.

Some reptiles lay their eggs all at one time, others may lay several times during a protracted breeding season. In cold regions females may reproduce only every other year. In general, reptiles lay fewer eggs than amphibians. The largest clutches, up to 200, have been reported for some of the marine turtles. It is noteworthy that these are the forms whose egg-laying area is most restricted. On the other hand, some lizards and snakes produce only one or two eggs at a time. Older females are apt to lay more eggs than younger ones, and larger species more than smaller ones.

Structure of Egg and Embryo

The egg and developing embryo of the American Alligator may be used both to illustrate the typical amniote shelled egg, characteristic of the reptiles, and to point out some of the ways the reptilian egg differs from the more familiar hen's egg.

The eggs vary in size, ranging from 65 to 85 mm. in their longest diameter and from 38 to 50 mm. in their shortest. At first they are slimy and pure white in color, but they generally become stained after a time by the damp, decaying vegetation of which the nest is built. The shell is thicker and coarser in texture than that of the hen's egg. It is quite calcareous and is readily dissolved in dilute acids. Inside the shell are two indistinct layers of shell membranes. No air chamber, like that of the hen's egg, forms between them at any stage in development.

Shortly after it is laid, a typical alligator egg develops a white, chalky band around the lesser circumference. Usually it runs entirely around the egg, but sometimes only part way around. It varies in width from about 15 to as much as 35 mm. When the shell is removed, the band is found to be caused, not by a change in the shell itself, but by the development of an area of chalky substance in the shell membranes. This band is thought to aid in the passage of gases to and from the developing embryo.

The white of an alligator's egg is unusually dense as compared to the albumen of a hen's egg. The entire egg may be emptied from the shell and passed from one hand to the other without serious danger of rupturing either the mass of albumen or the enclosed yolk. The albumen is apparently not divided into layers of decreasing density as in the hen's egg. Its color varies, in different eggs, from a pale yellowish-white (the most usual color) to a decided green. The albumen of the crocodile is said to be normally light green in color.

The yolk is a pale yellow, spherical mass lying in the center of the white. Its diameter is so great that it comes very close to the shell around the small circumference of the egg where it is covered only by a thin layer of white. Care must be taken not to rupture the yolk in removing the shell from this portion. The yolk substance is quite fluid and is contained in a rather delicate vitelline membrane.

Development starts in the oviduct and by the time the egg is laid a definite young embryo has formed. The amnion develops rapidly and entirely from the anterior end of the embryo. The relationship between embryo, yolk mass, amnion, and chorion is illustrated in Figure 3-1. Like the embryo of the chicken, the alligator embryo shows torsion or twisting of the body, so that the anterior end lies on its side, but unlike the chicken, which always twists to the right, the alligator embryo may turn either to the left or to the right.

Just before hatching, the young alligator is about 20 cm. long, nearly three times the length of the egg, but the compressed tail is bent so that, although it forms about half the length of the animal, it really takes up little room within the egg. The umbilical cord is withdrawn, but the scar is still present at the time of hatching.

Parental Care

Most reptiles abandon their eggs as soon as they are laid. Indeed, the mother may never even see them. Reports of a female guarding her clutch, based simply on the chance observation of an individual in the vicinity of the eggs, should be viewed with skepticism. Nevertheless, there is evidence that at least a few reptiles have developed care-giving behavior, beyond that shown in the preparation of a proper incubation site. Female skinks of the genus *Eumeces* stay with the eggs and protect them from small predators. The mother may turn the eggs periodically and bring them together if the clutch is accidentally scattered. It has even been suggested that she helps to incubate them by basking awhile, then returning to coil her sun-warmed body around them.

Females of several species of snakes are known to coil about the incubating eggs. Perhaps best known are the Pythons. After depositing the eggs, the female draws them together by movements of her tail and body. She heaps them in a pyramidal mass and coils around them, with her head capping the spiral formed by her body. She stays with the eggs for six weeks, only rarely moving away to drink, but leaves them about two weeks before they hatch. Reports that the temperature of the female's body is raised during incubation, as it is in the birds, are conflicting and in need of further confirmation. Other snakes that are known to guard the eggs include the cobras and the American Mud Snake (*Farancia abacura*).

The best-developed care-giving behavior pattern so far reported for a reptile is that of the American Alligator. During the incubation period of nine or ten weeks, the mother remains in the vicinity of the nest, and is said to moisten it occasionally with the contents of her bladder. When the young hatch, they start a high-pitched grunting which attracts the attention of the mother. If the surface of the nest is packed too hard, she may scrape it away to release the young. She is said to scoop out a wallow pool for them and to guard them from attack by predators, including other alligators. The young may remain with the mother for a year or more.

One of the most prevalent and persistent of all snake myths is the story of the mother swallowing her young to protect them and later disgorging them. The tale is usually told of live-bearing species. It is untrue, of course, but it may have this much basis in fact. In captivity the young of some ovoviviparous species (for example, the European Viper) have been seen to remain in the vicinity of the mother for a couple of days after birth and to seek shelter beneath her body when threatened. Whether they do the same in the wild is not certain. Such an association must be at best an ephemeral one.

HATCHING, BIRTH, AND YOUNG

Heat is necessary for the development of vertebrate embryos and the rate of development is, in part, determined by the amount of heat available. The endothermic birds and mammals provide a relatively high and constant source of heat for their developing young, either by retaining them in the body of the mother, or by setting on the eggs. The length of the embryonic period is fairly constant for each species, though it varies, of course, from one species to another. In contrast, eggs of the ectothermic reptiles, like their parents, are dependent on environmental heat, and since this may vary widely from year to year and from place to place, the length of time it takes reptile eggs to hatch, even within a single species, is extremely variable. The hotter the summer, the sooner the young reptiles appear. In France, young of the Smooth Snake (*Coronella austriaca*) usually hatch in late August or early September, but if the summer is cold they may not appear until October. In the northern United States and Canada, young Painted Turtles (*Chrysemys picta*) may winter in the eggs and hatch the following spring.

Even in live-bearing reptiles, the gestation period is variable. Pregnant females spend much time basking and the young are born earlier in a warm summer when there has been plenty of sunshine. In England, young of the Slow-worm (a lizard, *Anguis fragilis*) and of the Viper (*Vipera berus*) may remain in the body of the mother over the winter following a cool summer.

Table 7-3 gives some indication of the range of incubation periods that have been reported for certain species.

Eggs of the Tuatara (*Sphenodon punctatus*) are said to take thirteen months to develop.

A young reptile ready to hatch needs some means of cutting its way out of the egg. Embryonic turtles and crocodilians have a horny projection called the caruncle on the tip of the snout with which they pierce or slash the shell. Lizards and snakes bear an egg tooth on the front of the premaxillary bone in the upper jaw. This egg tooth is shed within a day or two after birth, but the caruncle may persist for weeks. Young of ovoviviparous species are usually born encased in a thin membrane which can easily be ruptured by the snout; in them the egg tooth has degenerated. Sometimes the membrane is broken before the young leave the body of the mother. Once the first break in the shell is made, hatching usually proceeds slowly; it may take the young reptile a day or more to free itself completely from the shell.

Young reptiles resemble their parents in structure and in behavior. Baby

TABLE 7-3.   LENGTH OF INCUBATION PERIOD FOR VARIOUS
SPECIES OF REPTILES

Species	Incubation period
Caretta c. caretta	31 to 65 days
Chelodina longicollis	3 to 4 months
Chelydra serpentina	81 to 90 days
Eumeces fasciatus	4 to 7 weeks
Lacerta a. agilis	7 to 12 weeks
Sceloporus undulatus	10 to 12 weeks
Alsophis angulifer	89 to 97 days
Coluber constrictor mormon	51 to 64 days
Elaphe g. emoryi	72 to 77 days
Lampropeltis getulus californiae	66 to 83 days
Naja naja	69 to 76 days
Pituophis m. annectens	64 to 77 days
Pituophis m. melanoleucus	59 to 101 days

vipers and crotalids have well-developed and functional fangs, baby cobras can spread their hoods. Within a day or two most young reptiles seek the small prey that surrounds them. The skin is first shed shortly after birth.

FIG. 7-4. Hatchling Mountain Patch-nosed Snakes, *Salvadora grahamiae*, as they emerge from the eggs.

Similar as they are to the adults in morphology, the young often differ strikingly from their parents in color. They are frequently much more brightly marked, and may have entirely different color patterns. Adult Black Racers (*Coluber c. constrictor*) are a smooth, velvety black above, but the young are vividly marked with reddish saddles and spots on a tan background.

GROWTH, MATURITY, AND LONGEVITY

The young grow rapidly at first, then more slowly. Some reptiles apparently reach a definite size and stop growing, others may grow, although very slowly in the later years, throughout their lives. As human children do, reptiles seem to grow in spurts, with periods of active growth alternating with periods in which growth is very slow. During hibernation growth practically stops.

Reptiles are ready to breed before they are full-grown. For some, at least, sexual maturity seems to be more a matter of reaching a certain size than a certain age. Males of the Red-eared Turtle (*Pseudemys scripta elegans*) are sexually mature when the plastron is 9 to 10 cm. long, but it may take individuals from 2 to 5 years to reach this size. Females mature at a plastral length of 15 to 19.5 cm., attained in from 3 to 8 years. American Alligators do not breed until they are nearly 2 m. long and 6 to 7 years old. Many lizards and snakes take about three years to reach sexual maturity. There is some indication that tropical forms, whose growth is not interrupted by periods of hibernation, mature more quickly. Among the Javanese snakes, the Blunthead (*Pareas carinatus*) is ready to breed in about 11 months, the Red-necked Keelback (*Rhabdophis subminiata*) in 13 months, and the Greater Rat Snake (*Ptyas mucosa*) in 20 months.

As with amphibians, our knowledge of the maximum age to which reptiles live is based largely on specimens kept in captivity. Table 7-4 shows the longevity records for a number of species. Apparently turtles tend to live longer than snakes, and snakes longer than lizards. Again there seems to be a negative correlation between the degree of activity and the duration of life.

Our knowledge of the reproductive habits of many species of reptiles is either a complete blank or consists solely of a record of a single clutch of eggs discovered by chance. Yet it is clear that, in spite of the variations in detail, the overall pattern of reproduction is much less diversified for the reptiles than for the amphibians. Fertilization is always internal; if eggs are laid they are always laid on land; development is always direct.

The diversity shown by the modern reptiles is just that which must also

TABLE 7-4.   LONGEVITY OF SOME SPECIES OF REPTILES

Testudinata	Maximum age (years)
Chelydridae (*Macroclemys temmincki*)	59
Testudinidae (*Testudo sumeirei*)	152
Cheloniidae (*Caretta caretta*)	33
Trionychidae (*Trionyx triunguis*)	25
Pelomedusidae (*Pelusios derbianus*)	41 (left alive)
Chelidae (*Chelodina longicollis*)	37 (left alive)
Rhynchocephalia	
Sphenodontidae (*Sphenodon punctatus*)	28 (left alive)
Squamata **Lacertilia**	
Gekkonidae (*Tarentola mauritanica*)	7
Iguanidae (*Conolophus subcristatus*)	15
Agamidae (*Physignathus lesueuri*)	6
Scincidae (*Egernia cunninghami*)	20
Cordylidae (*Cordylus giganteus*)	5
Teiidae (*Tupinambis teguixin*)	13
Anguinidae (*Anguis fragilis*)	54
Helodermidae (*Heloderma suspectum*) ·	20
Varanidae (*Varanus varius*)	7
Serpentes	
Boidae (*Eunectes murinus*)	29
(*Constrictor constrictor*)	23
Colubridae (*Elaphe situla*)	23
Elapidae (*Naja melanoleuca*)	29
Viperidae (*Vipera ammodytes*)	22
(*Agkistrodon piscivorus*)	21
Crocodilia	
Crocodilidae (*Alligator mississippiensis*)	56

have been present in the ancestral forms to have allowed the evolution of the two great groups that arose from them. The birds specialized in the development of the hard-shelled egg, well supplied with albumen; the higher mammals in viviparity.

Collateral Reading and General Reference

Angel, F. *Vie et Moeurs des Serpentes*. Paris: Payot, 1950. (A natural history of snakes, cosmopolitan in scope.)

Gadow, H. *Amphibia and Reptiles*. Cambridge Natural History, vol. 8, London, 1909.

McIlhenny, E. A. *The Alligator's Life History*. Boston: Christopher Publishing House, 1935. (Based on first-hand experience with our most spectacular native reptile.)

Mertens, R. *The World of Amphibians and Reptiles*. Translated by H. W. Parker. New York: McGraw-Hill, 1960.

Oliver, J. A. *North American Amphibians and Reptiles*. Princeton, New Jersey: Van Nostrand, 1955.

Reese, A. M. *The Alligator and Its Allies*. New York: Putnam's, 1915.

Schmidt, K. P. and R. F. Inger. *Living Reptiles of the World*. Garden City, New York: Hanover House, 1957. (A beautifully and profusely illustrated account of the families of reptiles, including much life history data.)

Smith, M. *Fauna of British India*. Vol. I–III. London: Taylor and Francis, 1931–1943. (Contains much life history information.)
The British Amphibians and Reptiles. Rev. ed. London: Collins, 1954.

RELATION TO
ENVIRONMENT

AMPHIBIANS AND REPTILES, like all other organisms, live within an environment consisting of two parts: a series of physical factors, such as light, heat, air, and moisture; and a series of biotic factors, such as food, competitors, predators, and parasites. Together these determine the nature of the habitat of the animal.

The environment not only provides the means of existence for the animal —oxygen, water, food, and shelter—but it also imposes restraints upon the animal. It sets the bounds within which the animal must live and largely determines how many of each different kind will survive. Every species has an inherent power of reproducing itself, a biotic potential, far in excess of the number that actually can or do survive. The existing population of a species at any one time reflects a balance between its biotic potential and the resistance of the environment. When both are in equilibrium, just enough young reach maturity each year to replace the adults that die and thus the population remains about constant. The restraining factors of the environment themselves fluctuate. One year may be too wet, another too dry, one year food may be abundant, another year predators may be overly numerous. As a result, populations of animals also fluctuate. If environmental resistance is reduced, the population grows in number; if environmental resistance increases, the population becomes smaller. Generally these fluctuations tend to cancel each other so that over the years the numbers of a given species remain fairly constant. Drastic or long-term changes in the environment, however, produce permanent changes in populations. European settlers, with their guns and their custom of draining swamps to make

126

farmland, materially changed the environment of the American Alligators and permanently reduced their numbers.

The study of ecology is largely a study of the factors involved in environmental resistance. These factors are many, complex, and interrelated, but they are susceptible to analysis. From such studies certain basic principles have emerged. In 1840 Liebig first clearly expressed what is now known as Liebig's Law of the Minimum. This simply says that the factor present in minimum quantity is the one that offers the greatest restraint to a population. For example, if the food of a given species is very scarce in a region, that scarcity determines how many of that species can survive in that region. If food is plentiful but home sites are few, it is the scarcity of available home sites that limits the size of the population.

Liebig's Law has been modified and expanded through the years. Shelford has pointed out that it is not always a minimal factor that limits the distribution of an animal, sometimes it is a maximal factor. Too high temperatures, too much water, too much light, too many predators or parasites, restrain a population just as effectively as does too little of something. Thus Shelford expanded the Law of the Minimum into the Law of Tolerance which says that a species has a range of tolerance for a given environmental factor, bounded on one side by a minimum and on the other side by a maximum which set the limits of tolerance for that species. An animal may have a wide range of tolerance for one factor and a narrow range for another. Those animals with the widest ranges of tolerance for all factors are the ones most likely to be widely distributed. The period of reproduction and development is apt to be the critical one, for many environmental factors are more limiting on eggs, embryos, and larvae than they are on adults.

PHYSICAL FACTORS

Light

Some herptiles (most amphibians, geckos, some snakes) are nocturnal; others (most lizards, tortoises, some snakes) are diurnal. The amount of light varies between day and night, of course, but it is not the only factor that does so. Humidity is higher, temperature is lower, different prey and predator species are active at night. The nocturnal-diurnal activity cycle of an animal may be a response to one or more of these factors rather than directly to light. Green Treefrogs (*Hyla cinerea*) rest during the day and stir abroad at night. The resting sites they choose, however, are frequently exposed to full daylight and at night they are attracted to lighted windows as are the night-flying insects on which they feed. Here the activity cycle of the frogs may be determined by that of the insects.

Heat rays as well as light rays are received from the sun, so that the responses of animals in nature to sunlight may really be responses to heat. Controlled experiments, in which other factors are kept constant and only the amount of light is varied, do indicate that light alone can have an effect on herptiles. One such effect is simply to change the color of the individual. Kept at constant temperature and pressure, the Carolina Anole (*Anolis carolinensis*) turns green in the dark and brown in light, although there is some variation in the intensity needed to bring about a brown coloration in different individuals. We once observed a Green Treefrog (*Hyla cinerea*) resting with half of its body in sunlight and the other half in shade. The part in the light was bright apple green, the other part a dark olive. Tail fins of tadpoles of the Barking Treefrog (*Hyla gratiosa*) are transparent and light yellowish during the day, coal black at night. Other factors, such as temperature, humidity, and the emotional state of the individual, also influence color; hence the color of an animal in the field is probably the result of a combination of factors rather than of light alone.

Many animals that are nocturnal or crepuscular (active in the dim light of twilight or dawn) are thought to be simply negatively phototropic, that is, they tend to turn away from a source of light, but the light they avoid need not be the light that we see. The Two-toed Amphiuma (*Amphiuma m. means*) avoids direct light. It has been shown that ultraviolet rays injure the skin of this salamander. When light is filtered to shut out these rays the animal is not so strongly negatively phototropic. Limbless lizards of the genus *Anniella* are subterranean and crepuscular. Specimens kept in a laboratory where the burrows could be flooded with strong light through the plate glass of the terrarium showed no tendency to turn away from the light and indeed frequently remained out of the burrows in the bright light of the room. Plate glass filters some of the nonvisible rays and it may be these rays that cause *Anniella* to avoid the light of day. The presence or abundance in a given area of a species susceptible to injury by light may thus be governed by the amount of available cover or ground suitable for burrowing.

Preliminary counts indicate that the aquatic salamander *Necturus* has a higher blood cell count at night than during the daytime. This is probably correlated with greater nocturnal activity rather than directly with the amount of light.

Heat

Heat is produced in the animal body as a by-product of metabolic activity. The respiratory and circulatory systems of the herptiles are less efficient than those of the birds and mammals in bringing an abundant supply of oxygen to the tissues for metabolic activity. Since their metabolic rate is consequently low, the herptiles do not produce enough internal heat to maintain body temperatures high enough for an optimum existence. They must absorb additional heat from some outside source, directly from the sun's rays, from the surrounding medium, or from the substratum on which they rest. In short, the herptiles are ectothermic.

To say that the herptiles absorb heat from the environment does not mean that their body temperatures are the same as those of the environment. The Squamata, at least, show a remarkable ability to control the level of body temperature during their periods of activity and may be considerably warmer than the surrounding air. An iguanid lizard, *Liolaemus m. multi-formis*, taken at a high altitude in southern Peru, had a cloacal temperature of 31 C., although the air temperature in the shade nearby was only 0 C. A lizard may bask in the sun until it is sufficiently warmed up, then, to prevent its body temperature from going too high, may move into a burrow or den, into water, or simply into a shady spot. A Sidewinder (*Crotalus cerastes*) was observed to maintain its temperature between the narrow limits of 31 and 32 C. by simply exposing more or less of its body to direct solar radiation while coiled at the mouth of its burrow. The ability of some herptiles to change color may possibly serve as a thermoregulatory mechanism, since a dark color absorbs heat rays, and a light one reflects them. Nocturnal forms can draw upon the solar heat absorbed by the substratum during the day.

The critical minimum temperature is the point at which an animal suffers from cold torpor to such an extent that it is helpless to escape from enemies or to act to raise its body temperature. The critical maximum is that temperature at which locomotor activity becomes disorganized and the animal loses its ability to escape from conditions that will lead to its death. They are ecological lethal points, since, although the temperatures alone may not kill, they hinder the animal from carrying on activities that would prevent its destruction.

Between the critical minimum and critical maximum, a reptile has an eccritic temperature to which it tends to adjust its body temperature by thermoregulation and at which it seems to carry on metabolic activities to the optimum extent. The Desert Spiny Lizard (*Sceloporus m. magister*) feeds well and seems superficially to thrive at a body temperature of approximately 30 C., but defecates frequently only at body temperatures of about 37 to 38 C. A higher temperature may be necessary to bring about peristalsis; therefore the lizard probably thrives best at the higher temperature. These eccritic or "preferred" temperatures are surprisingly close to the critical maxima, in most reptiles being only about 5 to 6 C. lower.

Studies on a group of diurnal lizards (*Sceloporus occidentalis, S. undulatus, Uta stansburiana,* and *Uma inornata*) indicate that the parietal eye (a rudimentary third eye found in the center of the forehead in *Sphenodon* and some lizards) helps to regulate the amount of exposure to sunlight.

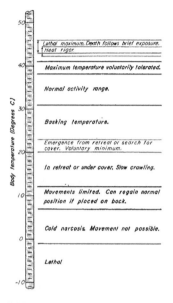

Lethal maximum. Death follows brief exposure.
Heat rigor.

Maximum temperature voluntarily tolerated.

Normal activity range.

Basking temperature.

Emergence from retreat or search for cover. Voluntary minimum.

In retreat or under cover. Slow crawling.

Movements limited. Can regain normal position if placed on back.

Cold narcosis. Movement not possible.

Lethal

Body temperature (Degrees C)

FIG. 8-1.

Significance of body temperature to the behavior of reptiles is indicated by this chart giving the approximate temperatures for various activities of Spiny Lizards (*Sceloporus*). The effects of exposure to heat levels near the extremes depend on duration. Even temperatures near the upper limit for the lizards' normal activity become lethal if exposure is prolonged. [After Bogert.]

Lizards in which the eye was removed or shielded by aluminum foil spent more time exposed to light on the surface of the ground than did untreated animals. The eccritic temperatures were the same for both groups, but the treated lizards maintained the thermal levels necessary for normal activity for longer times during the day. They were less viable than the controls when deprived of food, probably because they used up their available energy faster.

In the California desert, diurnal lizards maintain body temperatures of about 37 C. Nocturnal lizards not only tolerate but actually prefer lower body temperatures. Their eccritic temperatures are around 29 to 30 C. Snake body temperatures are usually a few degrees lower than those of lizards, and again the nocturnal snakes have lower eccritic temperatures than do the diurnal ones. For example, the Racer (*Masticophis flagellum piceus*), a diurnal snake, has an average activity temperature of 33 C., whereas the nocturnal Glossy Snake (*Arizona elegans occidentalis*) of the same region, has a normal activity temperature of about 27 C.

Species within a genus have about the same normal temperature range even though they may be quite widely separated geographically and climatically, whereas two different genera may be quite dissimilar even in almost identical habitats. The Desert Spiny Lizard (*Sceloporus m. magister*) has body temperatures ranging from 31 to 38 C., but those of the Checkered Whiptail (*Cnemidophorus t. tessellatus*) from the same region range between 37 and 44 C. On the other hand, the Florida

NORMAL ACTIVITY RANGES

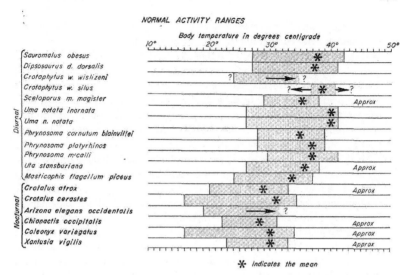

FIG. 8-2. Normal activity ranges of 18 reptiles of the southwestern United States, including both snakes and lizards. Twelve diurnal forms are arranged roughly in the order of size, with the larger species at the top of the column, and six nocturnal forms are arranged in the same fashion. Stippled portions of the bars indicate the normal activity range, that is, the temperatures between the minimum and maximum voluntarily tolerated by individual species under conditions (1) in the field, (2) in cages set up in the field, or (3) under laboratory conditions. The * indicates the mean for temperatures recorded when individuals were engaged in normal activities, and hence represents an approximation of the ecological optimum. Data for all species are not sufficiently extensive to be considered conclusive. Consequently portions of this chart are provisional, and revisions may prove necessary after controlled experiments in the laboratory have been conducted for the forms in question. [After Cowles and Bogert.]

Scrub Lizard (*Sceloporus woodi*) has body temperatures ranging from 31 to 39 C., and the Six-lined Racerunner (*Cnemidophorus sexlineatus*) from Florida ranges from 38 to 44 C. Thus the two species of *Sceloporus* have closely similar body temperatures and so do the two *Cnemidophorus*. It is evident that these lizards have the ability to regulate their thermal levels enough to keep them near their eccritic temperatures regardless of the habitat in which they are found.

The "coldest" reptile is the sluggish, nocturnal Tuatara (*Sphenodon*) found on cold, windy, foggy islands off the coast of New Zealand. Body temperatures of specimens taken in the wild at night ranged between 6 and 13.3 C. *Sphenodon* has an extremely low metabolic rate, and although it can move fairly rapidly for short distances, it is probably incapable of any sustained activity comparable to that of the lively lizards.

Male sex cells seem to be more sensitive to the effects of heat than other body tissues. Geographic variation in spermatogenic activity of the Garter Snakes (*Thamnophis*) may be correlated with seasonal increases and decreases in the number of hours a day the snakes can maintain their body temperatures within the normal activity range. Increase in temperature may cause sterility in the male Desert Night Lizard (*Xantusia vigilis*). Strangely enough, such correlations have not been found in the females.

Much has yet to be learned about amphibian body temperatures. Most species that have been tested show little control over their internal temperatures, but a few, such as the Italian *Salamandrina terdigitata*, do have some thermoregulatory ability. Active specimens of the Slimy Salamander (*Plethodon glutinosus*) of the eastern United States had body temperatures ranging from 16 to 20.3 C. A very low body temperature of 7.1 C. has been recorded for the Mount Lyell Salamander (*Hydromantes*) which lives at the borders of the snow caps. Salamanders that live in quiet ponds and streams and have lungs seem better able to stand high temperatures than do the lungless salamanders. Another factor besides temperature tolerance may be involved here. Cold water holds more dissolved oxygen than does warm water. Lungless salamanders depend on body surface absorption of water-dissolved oxygen but salamanders with lungs can come to the surface and gulp air. The latter may be able to live in warmer waters than the former, not only because they are more tolerant of high temperatures, but also because a supplemental source of oxygen is available to them.

Temperature undoubtedly plays an important role in determining the geographic distribution of the herptiles. They are limited to regions where sufficient environmental heat is available for their needs at least part of the year. The persistence of so many primitive species on islands and peninsulas may result in part from the equable maritime climate of these land forms. Islands do not present as rigorous an environment, so far as temperature fluctuations are concerned, as do the continental masses. The "temperate" climates of the large land masses, particularly outside of the tropics, are really most intemperate. In the later geologic epochs, these areas have been characterized by climatic extremes to which the primitive forms may have been unable to adjust. As a result they have been eliminated on the continents but have persisted on the islands.

Humidity

Amphibians have permeable skins through which water passes readily. How much moisture is lost through the skin depends in part on the humidity of the surrounding air and substratum. The more water vapor there is in the air (the higher the humidity) the less water will the air absorb from the animal's body. In very dry air moisture loss may soon reach a critical point. Amphibians are also able to take up water through their skins, but from the substratum rather than from the air. In studying the relations of amphibians to environmental moisture we must thus take into account the amount of moisture in the substratum as well as the humidity of the air.

Amphibians vary considerably in their ability to tolerate dehydration, and this seems to be definitely associated with habitat preference. Ten different kinds of frogs were tested in a dehydration chamber in which artificially dried air was passed over the animal being tested. The frogs were weighed before and after dehydration, with the amount of weight lost representing the amount of moisture lost during dehydration. Table 8-1 shows how much moisture could be lost without fatal results by individuals of these ten species, and the preferred habitat of each.

TABLE 8-1.   THE VITAL LIMITS OF WATER LOSS AND HABITATS OF FROGS (From Thorson and Svihla, 1943.)

Species	Pct. loss in body wt.	Habitat
Scaphiopus hammondi	48.8	Terrestro-fossorial
Scaphiopus holbrooki	48.1	Terrestro-fossorial
Bufo boreas	43.6	Terrestrial
Bufo terrestris	43.0	Terrestrial
Hyla regilla	39.0	Terrestrial
Hyla cinerea	37.3	Terrestro-arboreal
Rana pipiens	36.6	Terrestro-semiaquatic
Rana aurora	34.0	Semiaquatic
Rana sphenocephala	32.4	Semiaquatic
Rana grylio	31.2	Aquatic

This table shows that frogs occupying drier habitats tolerate dehydration better than those occupying more humid habitats. Spadefoots (Scaphiopus) are notably animals of dry land and can stand a loss of body weight of almost 50 percent. The most aquatic species on the list is the Pig Frog (Rana grylio), characteristically found sitting on lily pads or floating at the surface of the water in marshes of the southeastern United States. It could tolerate a loss of body weight of less than one-third.

Amphibians have various ways of coping with a reduced amount of environmental moisture. A Climbing Salamander, Aneides, coiled up in a

dehydration chamber, thereby reducing the amount of body surface exposed
to the dry air.

Many amphibians burrow in the ground to escape dry spells. Prolonged
seasonal droughts are characteristic of the climate of much of Australia. At
the approach of dry weather, frogs of the Australian species *Chiroleptes*
platycephalus inflate their bodies with water until they are practically round,
then burrow into the ground to wait for the return of the rains. Natives of
the region have an uncanny ability to locate the holes of these animals and,
if caught out in the desert without water, dig up the frogs and drink the
water stored in them.

Since most amphibians select habitats in which the amount of environmental moisture, both in the air and in the substratum, is below the maximum possible, it seems probable that too much environmental moisture can
be detrimental as well as too little. A dehydrated specimen of the salamander *Ensatina eschscholtzi*, when placed on a saturated substratum, stood up
so that only its feet touched the wet sand, a posture in which water absorption from the substratum was considerably reduced. These animals are
normally found in very humid surroundings where they may often encounter
excessive moisture in the ground. The posture may have represented a
genetically fixed response to a saturated substratum.

The cornified skin of a reptile is much less permeable than that of an

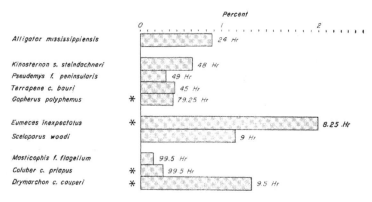

Percentages of initial body weight lost per hour by ten floridian reptiles (mean rate of loss
for the period of time indicated) under conditions of virtually constant temperature (38 C)
and relative humidity (37%). An ✳ indicates species that died in the period.

FIG. 8-3. Comparison of the rate of moisture loss (in terms of the percentage
of the initial body weight lost) of 10 species of reptiles from Florida under
constant conditions of temperature and relative humidity. [After Bogert and
Cowles.]

amphibian. Reptiles therefore lose water to the environment more slowly and have been able to adapt to drier habitats. Figure 8-3 compares the rate of moisture loss (expressed as percentage of initial body weight) of ten species of Florida reptiles. The Coachwhip (*Masticophis flagellum*) survived the longest in the thermal chamber and lost moisture at the slowest rate; of the species listed it has the widest range, being one of the few reptiles found from coast to coast in the United States. The Indigo Snake (*Drymarchon couperi*), which lost weight most rapidly, is more restricted in its geographic and ecologic range than either of the colubers. Found only on the moist southeastern coastal plain, it is most abundant in extreme southern Florida, where it shows a marked preference for a moist substratum. The Southeastern Five-lined Skink (*Eumeces inexpectatus*) also showed a high moisture loss; it is restricted to humid woodlands and does not inhabit dry, open country.

The results of these experiments on reptiles are not exactly comparable to those given above for frogs, since they deal with rate of dehydration rather than with total amount of moisture lost. The rate of moisture loss in the frogs did not show much correlation with tolerance of desiccation. Rate of dehydration does seem more important in the reptiles and may well be a limiting factor in ecologic and geographic distribution.

Salinity

Since amphibians evolved in fresh water, they are adjusted to the osmotic pressures of fresh water. This is particularly true for the eggs, which must absorb water from the surrounding medium in order to develop normally. Salts dissolved in water alter the osmotic pressure. Frog eggs immersed in only very slightly saline solutions do not absorb enough water for normal development. Sea water is thus an important barrier to the distribution of amphibians. Nevertheless, a few of them have been able to adapt themselves to waters with some degree of salinity. The European salamanders *Triturus vulgaris* and *Triturus helveticus* may breed in brackish interdune pools. Some of the toads (e.g., *Bufo bufo*, *Bufo viridis*) and some of the true frogs (e.g., *Rana temporaria* and *Rana cyanophlyctis*) are able to breed in more or less saline waters. At Manila Bay in the Philippines, *Rana moodei* apparently takes refuge in crab holes within the tidal zone. Some of these crab holes have masses of tadpoles in them and transforming tadpoles have been found in tidal pools with a salinity of 21 percent. The water over the crab holes has a salt concentration of 26 percent. We know nothing of the mechanism involved in the adaptation to salinity.

Reptiles have been able to adapt themselves to marine life much more readily than the amphibians. Best known of the present-day forms are the sea snakes and marine turtles. Not too far behind in adaptation to salt water are the Diamondback Terrapins (*Malaclemys*), many snakes of the genus *Natrix*, and some of the crocodilians. *Amblyrhynchus cristatus*, the Marine Iguana of the Galápagos Islands, feeds on marine algae at low tide. It has developed a strong, flattened swimming tail, but also retains powerful clawed limbs for landing on the shelving plates of lava that form the shore. The Javanese snake *Fordonia* lives in the burrows of marine crabs within the tidal zone.

No known reptile has eggs that can develop in salt water—or indeed in any water. All marine forms must either return to land to lay their eggs as do the marine turtles, or else bear living young as do some of the sea snakes.

Barometric Pressure

Air pressure drops as altitude increases and also varies with weather conditions at any given place. Changes in pressure are correlated with changes in such factors as temperature, humidity and available oxygen; it is thus difficult to isolate and study the effect of barometric pressure *per se*. Furthermore, very little work has been done on the reaction of the herptiles to changes in air pressure. However, during the winter in Florida, two species of tree frogs (*Hyla cinerea* and *H. squirella*) leave their customary daytime resting places and seek more sheltered spots shortly before sudden drops of temperature occur. These cold spells are initiated by the arrival of high pressure areas. It is possible that the movement of the frogs is triggered by the change in barometric pressure preceding the drop in temperature.

Oxygen–Carbon Dioxide Pressures

Since the relative proportions of the gases in the air remain quite constant at different heights, one of the most important results of the decreased air pressure at high altitudes is the decrease in the total amount of oxygen available. Hemoglobin is the substance in the red blood cells that carries oxygen from the lungs to the tissues. It is measured in parts per hundred by weight of the blood. To say that a frog has a hemoglobin value of 10 means that the weight of the hemoglobin makes up 10 percent of the total weight of the blood. Frogs from the lowlands have lower hemoglobin values than mountain forms. Hemoglobin values ranging from 7.5 to 13.5 have been reported for frogs and toads from the European lowlands. In the Alps, the toad *Bombina variegata* has values ranging from 14.25 to 15.25. In Guatemala, *Bufo marinus*, which occurs from sea level to elevations of about 1,500 m., has a mean hemoglobin value of 8.66, whereas *Bufo bocourti*, which ranges from about 1,700 m. to 3,400 m., has a mean value of 10.77. If the hemoglobins of these species have similar oxygen-carrying capacities (which is not yet known) the increase in hemoglobin could balance the decrease in oxygen at the higher altitudes.

Aquatic habitats differ greatly in the amounts of oxygen and carbon dioxide dissolved in the water. Cold water holds more oxygen than warm water. Green plants add oxygen to water and remove carbon dioxide during daylight hours. In standing water decaying vegetation collects on the bottom and since decay is an oxidizing process, free oxygen is removed from the water. If newts (*Diemictylus*) are placed in water and the oxygen tension is lowered, they come to the surface to gulp air or climb entirely out of the water. On the other hand, if oxygen tensions are maintained but the amount of carbon dioxide in the water is increased, the reaction of the newts is quite different. After about 11 minutes in water with an increased CO_2 content, the animals begin to display signs of discomfort and gape their mouths wide open under water. Soon they are unable to move about normally. They sink listlessly to the bottom and fail to respond to prodding. Raising the CO_2 content makes the water much more acid. The reaction of the newts may result from an upset in the physiology of respiration caused directly by the excessive CO_2 or it may result from the increased acidity of the water.

Acidity

Some natural waters are highly acid. Acidity seems to affect amphibian eggs much as does salinity, that is, it modifies osmotic pressure so that the eggs do not absorb enough water for normal development. Some amphibians do breed in ponds where the water is normally acid; their eggs are able to stand a higher degree of acidity than those of frogs that do not breed in such situations. Water from the sphagnum bogs of the pine barrens of New Jersey has a pH range from 3.6 to 5.2. Among the frogs found in the general area, Anderson's Treefrog (*Hyla andersoni*) and the Carpenter Frog (*Rana virgatipes*), the two forms most restricted to pine barren waters for breeding, have eggs that are able to develop normally at pH's as low as 3.8. Eggs of the Cricket Frog (*Acris gryllus*) are affected by a pH of 4.6; it does not breed in the waters of the pine barrens.

BIOTIC FACTORS

Since it is relatively simple to set up a controlled experiment in which only a single physical factor is varied, the precise effects of many of these

factors can be determined more readily than can the effects of the less easily
manipulated biotic factors. But no animal lives in a biotic vacuum. Herptiles
are subject to restraints by the other living things around them—the pred-

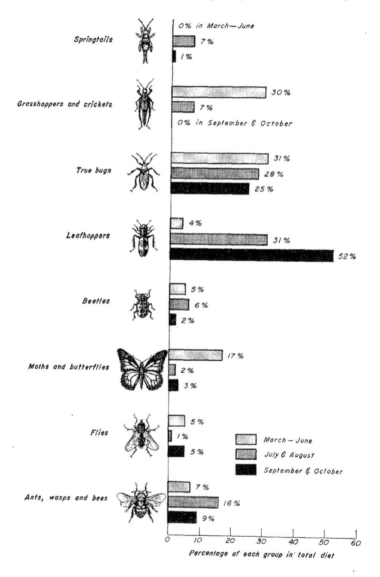

Percentage of each group in total diet

ators that prey upon them, the parasites that live at their expense, the competitors that vie with them for the limited necessities of their existence. Many of these interactions are superficially obvious, though often difficult to study quantitively. Others are less direct and less evident, consequently even more difficult to discover and evaluate.

Interactions between living things are frequently discussed under the term "biotic relationships." But since "relationship" has also the more restricted meaning of closeness of descent from a common ancestor, it will be less confusing if we call the situations discussed in this section "interactions" and retain the evolutionary implications of the word "relationship."

For convenience, we will divide these interactions into consort interactions, in which there is direct contact between the two organisms, and nonconsort interactions, in which one form affects another merely by living in the same community.

Consort Interactions

Food. Like all animals, the herptiles are dependent on other living things as sources of food. Usually they are carnivorous. Most amphibians and many reptiles feed on a wide variety of animals. Analysis of the stomach contents of some Greenhouse Frogs (*Eleutherodactylus*) showed that these little frogs had eaten representatives of eight different orders of insects, besides spiders, mites, centipedes, millipedes, and earthworms.

Turtles and crocodilians, which are able to tear off chunks of food, can feed on carrion or anything small enough for them to kill. A salamander, *Plethodon jordani*, will seize a large earthworm in its jaws and rotate its body rapidly until the worm breaks. It then swallows the fragment before attacking the remainder. Most herptiles, though, must swallow their food whole. For them, size is an important factor in determining selection of food. Growth of individuals of a prey species may remove them from among the items suitable for the predator. Early in the spring, the Northern Side-blotched Uta (*Uta s. stansburiana*) feeds mainly on young grasshoppers,

---

FIG. 8-4. Seasonal change in food composition of the Northern Side-blotched Uta (*Uta stansburiana stansburiana*) in Utah. Percentages are those of total number of items represented. Note particularly the increase in the leafhoppers consumed as the season progresses. This is due to the eating of large numbers of the Beet Leafhopper (*Eutettix tenellus*) as they become abundant. [Data from George F. Knowlton.]

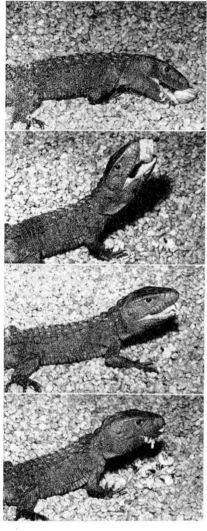

FIG. 8-5.
The Caiman Lizard, *Dracaena guianensis*, feeding on a snail.

but in the summer the grass-hoppers have grown rather large for this small lizard to handle and it shifts to bugs and leaf hoppers.

Conversely, many herptiles themselves change their diets as they grow. Young of the Suwannee Turtle (*Pseudemys floridana suwanniensis*) are largely carnivorous, feeding on ant larvae and other small invertebrates. The herbivorous adults graze on flats of eel-grass near the mouth of the Suwannee River.

Some reptiles have developed highly specialized food habits. Snakes of the genus *Dipsas* in South America prey exclusively on snails, and snails are also important in the diet of the Caiman Lizard (*Dracaena guianensis*).

Herptiles vary greatly in the amount eaten and the intervals at which food is taken. Some, such as the toads, eat large meals daily. A toad, indeed, may keep on eating even when it is stuffed to capacity. A large female American Toad (*Bufo americanus*) ate 152 Mexican Bean Beetles in a day. It has been calculated that from May through September a single toad could possibly eat 22,700 of these destructive insects. This is but one example of an amphibian that is beneficial to man by its feeding habits.

At the other extreme from the toads we find the snakes, which eat much less often. This is probably correlated to a large extent with the snake's ability to distend its jaws and thus swallow food items of great size. These make a much more lasting meal than the tiny items taken by frogs and lizards.

Many turtles and some lizards depend largely on plants as food. Of the big Land Iguanas (*Conolophus*) of the Galápagos Islands, Darwin found to his surprise that "to obtain acacia leaves they crawl up low stunted trees. It is not uncommon to see a pair quietly browsing . . . on a branch several feet above the ground." He amused himself by tossing a branch among a group of them and watching them attempt to seize and carry it off "like so many hungry dogs with a bone." Shingleback Lizards of Australia (*Trachysaurus rugosa*) eat berries and toadstools. When these large skinks live close to settled districts, they may be serious pests. They are particularly destructive to strawberries, and can devastate a crop by eating the unripe berries.

Predation. Although many herptiles are predators, most of them are also small enough to form the prey of many other animals. The tree frog eats insects and the snake eats the frog. Whether we regard the frog as predator or prey depends on our point of view.

It is doubtful if anyone could spend much time collecting without becoming aware of the importance of predation in the lives of amphibians and reptiles. An evening spent by a pond where a frog chorus is in full swing is enough to show that water snakes have been attracted to the same spot and are busily feeding on the frogs. Predation is not usually deleterious to a species though it is, of course, destructive to the individual. Studies on other animal groups have shown that a natural balance in numbers is maintained between the predator and prey species. When predators are removed, a population may become so numerous as to destroy or at least seriously deplete its own food supply; then starvation reduces the population to a level perhaps lower than it maintained when predators were present. Also, since predators catch the animals most easily caught, they remove from a population a large proportion of the sick and the unfit. They may check the spread of diseases and help maintain genetic fitness by eliminating the carriers of deleterious genes. These interactions have been most thoroughly studied among birds and mammals. Undoubtedly they apply to reptiles and amphibians as well, but there is a great need for quantitative studies and critical analyses of the effects of predation on the herptiles.

Studies on two species of Spiny Lizards (*Sceloporus graciosus* and *S. occidentalis*) showed that about 65 percent of the adult population of *S. graciosus* was replaced after an interval of three years, whereas there was 80 percent or more replacement of *S. occidentalis* after a single year's lapse. This was attributed in part to differences in predation. In the habitat of *graciosus* at the locality of the study, only one snake predator, *Thamnophis sirtalis*, is present. Six known or presumed snake predators (*Coluber constrictor, Lampropeltis zonata, L. getulus, Pituophis catenifer, Thamnophis ordinoides*, and *Crotalus viridis*) are known to live in the same general habitat as *Sceloporus occidentalis*.

A useful index to the extent of predation is the amount of tail breakage in lizards. In the population of the lizard, *Cnemidophorus lemniscatus*, on the Bay Islands of Honduras, only one-eighth as many individuals have broken tails as in the population on the mainland. There are very few predators on the Bay Islands, but many on the mainland. The Bay Islands population is also much more variable than the mainland population. This increased variability may result simply from decreased natural selection. Many of the variants that are eliminated by predators on the mainland may survive and propagate on the islands.

Symbiosis. This is a special and close association between two different kinds of organisms. Literally, they "live together." Symbiotic associations are classified according to whether both members benefit (mutualism), one member benefits and the other is unaffected (commensalism), or one member benefits and the other is injured (parasitism). The various categories grade into one another so that it is often difficult to say how we should classify a given association. We are often handicapped, too, by lack of knowledge. We know that two forms are closely associated, but we do not know how they affect one another. The categories should be regarded, not as hard and fast divisions, but as convenient groupings for the discussion of a variety of situations.

Mutualism. Only a few instances of true mutualism have been reported among the herptiles. A species of green alga grows inside the outer membranes of the eggs of the Spotted Salamander (*Ambystoma maculatum*). This is apparently beneficial to both organisms, for eggs inhabited by algae produce larger embryos, hatch earlier, and have a lower mortality rate than those without them. And the algae seem to grow more vigorously in eggs containing embryos than in those from which the embryos have been removed. The mechanisms by which plant and embryo affect each other are unknown.

Some aquatic turtles are real mossbacks, their shells being covered with dense growths of algae. The turtle provides a place of attachment for the plants, and the plants help conceal the turtle as it lies quietly on the bottom.

One type of mutualism is "cleaning" in which an organism removes, and eats, deleterious organisms from the body of another. On the Galápagos Islands, large red crabs have been seen crawling over the bodies of Marine Iguanas (*Amblyrhynchus*) and pulling ticks from their hides. Nearly 2,500 years ago, the Greek historian, Herodotus, reported that crocodile birds enter the mouths of basking crocodiles to remove leeches from their gums. The story has never been proved, but if it is true it is undoubtedly the oldest record of this type of symbiosis.

T. W. Kirkpatrick, in his book *Insect Life in the Tropics*, reports a situation that strongly suggests genuine mutualism and that would well repay further investigation.

"In East Africa a 'blind snake' (although actually it has vestigial eyes) of the genus *Typhlops* [now considered a legless lizard] lives with the safari ants, *Dorylus*, and accompanies them when they change their nesting site. I have several times seen one of these short thick snakes near the head of a marching column, with a few ants, apparently path-finders, going ahead and a number riding on its back. The main column carrying the brood followed behind and one got the impression that the snake was being employed as a bulldozer. Nothing is known about the biology of these snakes, though they are probably scavengers in the ants' nest. It is curious that I have never seen more than a single snake accompanying a column of ants, for if they bred in the ants' nest one would expect on occasions to find more than one. An East African native told me that in order to destroy a colony of 'siafu' it was only necessary to kill their tame snake, but it seems unlikely that this should be correct."

Commensalism. This is a one-sided symbiotic interaction in which one member derives some advantage from the association—protection, transportation, or a share of the other's food supply—and the other is neither injured nor benefited. One kind of commensalism is inquilinism in which a species lives in the domicile constructed and still occupied by another. Herptiles may be guests of, or hosts to, many other animals. Perhaps the best known inquilines are the ones that live with the gopher tortoises. Sidewinders (*Crotalus cerastes*), Great Basin Rattlesnakes (*Crotalus viridis*), Spotted Night Snakes (*Hypsiglena torquata*), and Banded Geckos (*Coleonyx variegatus*) have all been found hibernating in dens of the Desert Tortoise (*Gopherus agassizi*). Occasionally Diamondback Rattlesnakes (*Crotalus adamanteus*) and typically Gopher Frogs (*Rana capito*) live in the burrows of the Southeastern Gopher Tortoise (*Gopherus polyphemus*).

The Tuatara (*Sphenodon*) often makes its home in the burrow of a bird, the Sooty Shearwater.

Many different kinds of protozoans are present in the intestines of herptiles. They are usually considered parasites, but frequently they seem to do no appreciable injury to the host and perhaps should be classed as commensals. It is even possible that some of them are beneficial.

Parasitism. Parasitism is a special type of symbiotic association in which an organism spends most or all of its life on or in the body of another, derives its food from its host, and frequently damages the tissues of the host. Parasitologists have found the herptiles fruitful sources of parasites, both external and internal. Mites, ticks, leeches, copepods, fly larvae, flukes, tapeworms, threadworms, thorny-headed worms, and protozoans have all been reported to be parasites of the herptiles.

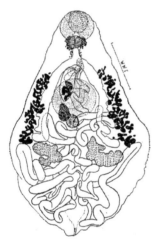

Outside of that plague of the zoo keeper, the mites, perhaps the best known parasites of reptiles are the flukes (trematodes) found in the respiratory and digestive systems of snakes. Most snake autopsies show heavy infections of these trematodes, but such infestations are not generally fatal. Indeed, the presence of trematodes does not seem to have much effect on the well-being of the snakes. Studies on the poisonous *Denisonia* and *Notechis* of Australia suggest that parasites may diminish their venom yields.

FIG. 8-6.

A Lung Fluke, *Pneumatophilus leidyi*, from the Banded Water Snake, *Natrix sipedon fasciata.* [After Byrd and Denton.]

There are some interesting correlations between trematode infections and the habits of snakes. Two species of Garter Snakes (*Thamnophis sirtalis* and *T. butleri*) differ quite pronouncedly; the average specimen of *T. sirtalis* is heavily infected with trematodes, *T. butleri* seldom shows them. These parasites pass through the early stages of their development in the body of a snail. After they leave the snail host, the larvae are ingested by an amphibian in whose body they undergo further development. If the amphibian is then eaten by a snake, the snake becomes infected and the trematodes here reach adult form. Since *T. sirtalis* feeds largely on amphibians, it is usually heavily parasitized. *T. butleri* eats earthworms and leeches rather than amphibians and is thus protected from the flukes. The presence of the parasite here provides a clue to the diet of the host.

For an animal species and its parasites both to survive, there must be an adjustment between them. If the members of a parasitic species are so numerous and so deadly as to kill off all available hosts, then that species must itself become extinct. Parasitism therefore tends to evolve toward commensalism. Adjustment between the trematodes and snakes seems to be on the individual level, that is, the body of the individual snake appears able to withstand the damage done by the parasites. Adjustment may also be on the population level. In the vicinity of Gainesville, Florida, Carolina Anoles (*Anolis carolinensis*) are parasitized by the larvae of a fly of the family Sarcophagidae. In some years, during May and June the bodies of dying anoles are found on the ground, their insides literally eaten out by maggots which can be seen emerging through a hole in the body wall. As many as 34 maggots have been taken from a single anole. During other years no evidence of such parasitism is found. It is probable that the periods of heavy infestation follow periods of peak abundance in the anoles and that the numbers of the two populations fluctuate together so that there are never enough flies to reduce the number of lizards below the level necessary to maintain the population.

One of the most interesting studies of amphibian parasitology was the attempt by Dr. Maynard M. Metcalf to correlate the evolution of the ciliate protozoans of the family Opalinidae with the evolution of the frogs they parasitize. Since opalinids are not strictly host-specific, but apparently sometimes transfer from one family to another, they do not reveal much about the evolution of the frogs. The main interest of Metcalf's work lies in the method of gathering and combining data from the organisms and from their parasites. Since host and parasite do evolve together, this method should prove useful in future evolutionary studies.

Nonconsort Interactions

The interactions discussed above all involve direct association between the individuals of one species and those of another species. There are other interactions between species that have a bearing on the well-being of the animals although there is no direct contact between individuals.

Habitat Formation. Plants play a most important role in the general environment of the herptiles. They make up a major part of the habitats in which most species live and largely determine the visible characteristics of

these habitats. They have a profound effect on the physical features of an
environment. In forested regions they serve as buffers against the extremes
of heat and cold found in deserts, their roots hold moisture in the soil, the
water they transpire raises the humidity of the air, they provide shade,
and they contribute to the formation of soils. Rotten logs form protective
retreats for many of the herptiles; some terrestrial salamanders may spend
most of their lives within or under a single large log. The leaf mold on
the ground provides a habitat for small salamanders, frogs, and snakes.
Many herptiles lay their eggs or bear their young in rotten logs, dead
stumps, or leaf mold.

Buffer Species. Studies of the feeding habits of otters in Michigan trout
streams showed that they ate 25.9 percent forage fish, 25.3 percent amphibians, and 22.7 percent game and pan fish. If the number of forage fish
should be reduced, the otters would eat more amphibians and game fish.
This is what we call buffering, that is, the amphibians and trout are protected from excessive otter predation by the presence of the forage fish
and the amphibians in turn help relieve predation on the game fish. The
number of amphibians present in the area is in part determined by the
number of forage fish, and also helps determine the number of game fish,
although there is no direct contact between them. Since populations of
many, if not all, animal species alternate between periods of abundance and
scarcity, and since many predators show few specific preferences in food,
simply eating what is most available (usually the most abundant species),
the buffering interaction is apt to be fluctuating and reciprocal. If members
of one species are very common in a given year, predators will eat many of
them and a less common species will be buffered. The next year numbers
of the first species may be reduced, in part because of heavy predation.
The second species, which was protected the first year, may become the
commoner one, and the predators will turn their attention to it. So a species
that was buffered one year may be itself the buffer another year. So far
as herptiles are concerned, these interactions are largely theoretical. But it
is logical to assume that since they have been reported for other animals,
they probably also occur in herptiles. We need long continued studies of
predation, correlated with studies of the relative abundance of prey species,
before we can say definitely that they do occur.

Mimicry. Many dangerous or distasteful animals have striking color patterns which notify potential predators that these animals had best be let alone.
Coral snakes (*Micrurus* and *Micruroides*) display such warnings with bright
red, yellow or white, and black rings that encircle their bodies. In regions
where these snakes are found, other harmless snakes (e.g., *Atractus latifrons,*

A. elaps, Procinura aemula, Lampropeltis doliata) have similar patterns and thus share in the protection afforded by the warning coloration of the coral snakes. This is mimicry, the result of convergent evolution brought about by natural selection. The harmless species is called the mimic and the one copied is called the model.

Competition. The interactions discussed above mainly involve animals that are not closely related. Competition, on the other hand, is apt to involve members of the same or of very similar species. This is because the more closely two forms are related, the more similar their ecological requirements are apt to be, and the more intense the competition between them.

Many closely related species are allopatric, that is, they live in different areas. Where two closely related species are sympatric (living in the same general area) they almost invariably develop slightly different ecological requirements so that competition between them is reduced. When two species with closely similar requirements do come to inhabit the same area, such overlapping is usually only temporary. One or the other will be eliminated by competition.

Three species of Garter Snakes occurring in Michigan—*Thamnophis sirtalis*, the Common Garter Snake, *T. sauritus*, the Ribbon Snake, and *T. butleri*, Butler's Garter Snake—are found together in high concentrations near the hibernating sites in the spring and fall. The Common Garter Snake and the Ribbon Snake are about the same size but differ greatly in feeding habits. A recent study showed that amphibians made up 90 percent of the food of the former but only 15 percent of the food of the latter. Earthworms formed 80 percent of the food of the Ribbon Snake and 83 percent of the food of the smaller Butler's Garter Snake. The latter did not eat amphibians but 10 percent of its food was leeches. During the summer months the snakes dispersed. Butler's Garter Snakes were found most often in grassy areas near water. Ribbon Snakes were also found near water, but in bushy rather than grassy areas. Common Garter Snakes were much less restricted in habitat. They overlapped the other two in distribution and were also found in areas where the others were not usually seen. Thus the two species most similar in food preference, *T. butleri* and *T. sauritus*, live in different minor habitats during the summer months when they are feeding most actively. The species (*T. sirtalis*) that overlaps the other two in habitat distribution differs greatly from both in food eaten. Competition between the three is reduced and they are able to live side by side.

Since large animals tend to eat larger food items than small animals, differences in size may reduce competition between two similar species sufficiently so that they are able to coexist in the same habitat. In Jamaica, three closely related species of frogs of the genus *Eleutherodactylus* may occur in the same place; a large *E. pantoni*, a moderate-sized *E. gossei*, and a diminutive *E. andrewsi* may all live under the same rock. Probably the difference in size is enough to cause a difference in choice of food items so that the three are not really competitors.

Difference in size probably helps also to eliminate competition between young and adults of the same species. During the period of emergence, the young of a species are tremendously abundant as compared to the adults. Competition between the young must be very severe and it may be this that compels the young to move away from the breeding sites. There are indications that new territories are usually invaded by young frogs rather than by adults.

There are probably many other interactions between species about which we know little or nothing. For instance, it has been shown that of the total population of snakes in an area, about half of the individuals belong to about 10 percent of the species and about 10 percent of the individuals belong to half of the species. This holds true in Panama, with its tremendously large snake fauna, and in places like Pennsylvania and California where fewer species and fewer individuals are found. Why this is so is not known, but it is hardly likely that these ratios, occurring as they do in such distant geographic areas, are coincidental.

Collateral Reading and General Reference

Allee, W. C., A. E. Emerson, O. Park, T. Park, and K. P. Schmidt. *Principles of Animal Ecology*. Philadelphia and London: W. B. Saunders, 1949. (A standard, comprehensive work on animal ecology.)

Angel, F. *Vie et Moeurs des Amphibiens*. Paris: Payot, 1947.
Vie et Moeurs des Serpentes. Paris: Payot, 1950.

Bogert, C. M. and R. B. Cowles. "A Preliminary Study of the Thermal Requirements of Desert Reptiles." *Bulletin of the American Museum of Natural History*, vol. 83, article 5, 1944. (In this and other papers, Bogert and Cowles have contributed largely to our understanding of temperature as a factor in the environment of the herptiles.)

Noble, G. K. *The Biology of the Amphibia*. New York: McGraw-Hill, 1931. Reprinted by Dover Publications, 1954.

Smith, M. *The British Amphibians and Reptiles*. Rev. ed. London: Collins, 1954.

BEHAVIOR

IF WE APPROACH a small wood pond, on the bank of which a number of frogs are sitting quietly, we are greeted by a series of splashes as the frogs jump into the water and swim down to hide under the detritus at the bottom. They have been stimulated by the sight or sound of our approach and have responded with an escape reaction. In doing so they have exemplified one definition of behavior—a response to a stimulus.

A stimulus always involves some change in conditions, but the change need not be external. The sensation of hunger is an internal stimulus to which a snake responds by moving around, thereby increasing its chance of coming in contact with prey. The sight or scent of the prey then acts as an external stimulus to bring forth the striking response.

Furthermore, there seems to be some mechanism whereby a stimulus that does not evoke a reaction may yet affect the central nervous system so that later behavior is modified. Thus an animal may observe an object in the vicinity of its home without any reaction until it is returning from a distance. Then the sight of the familiar object causes the animal to turn in the direction of its home.

There are three approaches to the study of behavior. The neurologist studies the anatomy and physiology of the nervous system in an attempt to explain how nerve impulses originate, how they travel from one part of the system to another, how a stimulus is converted into a response. The animal psychologist studies the behavior of an animal under controlled conditions. He seeks to find out what stimulus brings forth what response and whether and how a given response may be modified by experience (learning). The naturalist observes the actual behavior of an animal under field conditions, how it finds food and mate and shelter. Progress has been made in all three of these fields, but there are still enormous gaps—gaps that must

be bridged before we can begin to present a unified and meaningful picture
of behavior.

STRUCTURAL BASIS OF BEHAVIOR

How an animal behaves is basically determined by its structural characteristics. Except for the simplest organisms, all animals have receptors,
sense organs that can be stimulated by changes in the environment. They
also have conductors, nerve pathways that relay the stimuli received by
the sense organs and converted by them into nerve impulses, to the effectors, either the muscles or the glands of the animal. The contraction of
these muscles or secretion of these glands produce responses to stimuli.

Nervous System

Interposed between the sensory receptors and the effectors is the nervous
system, comprised, in the vertebrates, of the nerves, spinal cord, and brain.
Each nerve cell (neuron) consists of a cell body and two or more processes.
Each sensory (afferent) neuron is connected by a long process (nerve fiber)
to a receptor, and each motor (efferent) neuron by a similar fiber to an
effector. The nerves that we see when we dissect an animal are made up
of bundles of many of these fibers. Each neuron also has one or more other
processes which are in contact with the processes or cell bodies of other
neurons. From these processes impulses are "fired" from one cell to another. Also in the central nervous system are neurons that are neither
sensory nor motor. They are associative or internuncial (messenger) cells,
which relay impulses from sensory to motor neurons and also from one part
of the central nervous system to another. The numerous impulse transfer
points (synapses) between the different neurons make possible the co-
ordination of bodily activity, since the message from a single stimulus can
be relayed to a number of different effectors. They also make for flexibility
of response, by opening a number of potential pathways to a nerve impulse.

We often speak of impulse transmission as though we were dealing with
a single series of changes following a single path from receptor to effector.
But this is oversimplification. Usually, though not always, each separate
stimulus initiates impulses along a number of sensory fibers. The eye, for
instance, contains the nerve endings of many neurons. It would be practically impossible to stimulate just a single one. But even a single impulse,
traveling along a single sensory fiber, seems to spread out when it reaches
the central nervous system, like water sprayed from a hose. Almost without

exception, a response is brought about by the contraction of at least several different muscles, each of which is activated by many different motor neurons. We are dealing more with areas of transmission than with simple cell-to-cell pathways. But it is easier to speak of an impulse as a unit following a definite path.

An impulse tends to cross more readily a synapse that has been crossed before. Thus the impulse set up by a given stimulus will usually follow the same course from cell to cell, and result in the same reaction, as it did before. In this way, pathways are established through the network of neurons and so habits are formed.

Many pathways are apparently predetermined. They are present at birth or develop with the maturation of the individual. They probably result from the spatial arrangement of the neurons and they give rise to behavior patterns that are stereotyped and do not have to be learned. The swimming of the larva of the salamander *Ambystoma* has been shown to be a function of the maturation of the animal. Of two lots of larvae of the same age, one was kept in natural water and the other in a solution of chlorotone, which inhibits activity but does not prevent growth and development. When the controls in natural water had developed the ability to swim, the other larvae were removed from the chlorotone and placed in natural water. Although they had not had the practice and experience of the free swimmers, in about 30 minutes they could swim just as well. This was not simply an accelerated period of learning, for when the larvae that had developed their swimming behavior in natural water were anesthetized with chlorotone, they also took about 30 minutes to resume normal swimming movements.

Most of the behavior of the herptiles is thus innate and stereotyped. Yet it is also true that no individual ever behaves exactly like another. It sometimes seems that the most striking characteristic of behavior is its variability from individual to individual and from time to time in the same individual. What are some of the possible reasons for this variability?

In the first place, it is probable that no two individuals have exactly the same number and arrangement of neurons in the central nervous system. Hence the potential pathways for an impulse may differ slightly from animal to animal.

The pathways established, the habits learned, are in part determined by the nature of the stimuli that the animal has been receiving from the beginning. No two individuals ever have exactly the same experiences. Two toads may transform in the same pond and move out on land together, but they cannot travel the same path at the same time, dig their burrows at the same spot, or catch and eat the same fly. The visual stimuli that signify the nearness of home to one may mean strange territory to another, and will evoke different responses.

Behavior may also be modified by changes in the chemistry of the blood. The effect of these changes is apparently to sensitize certain neurons so that the animal responds to stimuli that it ignores at other times. This is most clearly seen in the differences in behavior following the release of increased amounts of hormones with the onset of the breeding season. A female frog that is ready to lay responds to the sound of calling males by moving toward them; a spent female, or one that is immature, ignores the stimulus. Changes in blood sugar level also affect behavior. A hungry snake pays attention to a mouse in its cage; a snake that has recently fed may not.

Obviously, the more associative cells there are in the central nervous system, the greater the number of potential pathways, and the greater the possibility of variation in response. This is probably the reason that the reptiles show more complex behavior patterns than the amphibians. The cerebral hemispheres of amphibians are larger than those of fishes and the pallium is thicker, but this part of their brain is still concerned mainly with the reception and transmission of olfactory stimuli. The cerebral hemispheres of reptiles have increased markedly in size and the neopallium has appeared. Reptiles therefore show somewhat more complex behavior patterns, with a greater ability to learn, than the amphibians. But for all the herptiles, most behavior is under direct sensory control, and is largely stereotyped.

Sensory-Motor Basis of Behavior

In any study of behavior we must bear in mind that an animal is able to respond only to stimuli that it is able to perceive. If an animal cannot learn to distinguish between two differently colored cards, it may be color blind rather than stupid. And we must not make the mistake of assuming than an animal perceives only what we perceive. It may be reacting to stimuli of which we are not aware. Amphibians have chemical receptors scattered over the skin by which they can "feel" chemical differences in water that we can detect only by instruments or by using the special chemical receptors in our mouths and noses (tasting or smelling). Aquatic amphibians detect vibrations in the water through their lateral line organs and snakes hear vibrations passing through the ground. *Amphiuma* is sensitive to ultraviolet light (see p. 128). Among snakes, the pit vipers have a special sense organ located in a pit between the eye and nostril. This organ, which is sensitive to infrared rays, allows the animal to detect the proximity of warm-blooded prey by its difference in temperature from the surrounding air. So the first question in behavior studies is often "Just what stimuli is the animal capable of perceiving?"

The response an animal makes to a stimulus is also determined by its motor equipment. The true frogs (*Rana*) have long, strong hind legs. They are powerful jumpers and they usually sit so that one good jump will carry them to safety in the water. Their response to the approach of a large moving object (a possible predator) is to jump. Toads (*Bufo*) have shorter, less muscular legs. They cannot jump but can only hop. If a toad cannot quickly reach its burrow, it must rely on other means of defense, such as inflating its body, assuming a defensive pose, or secreting poison from its glands.

PATTERNS OF BEHAVIOR

We are still far from being able to explain much of the behavior of the herptiles in terms of stimulus-response mechanisms. This is partly because our knowledge of how herptiles behave under natural conditions is still incomplete. Since we cannot explain anything until we know what it is we are trying to explain, and since in this book we are primarily concerned anyhow with the animal in nature, we shall mainly confine ourselves in the rest of this chapter to a discussion of various patterns of herptile behavior.

Spatially Oriented Behavior

Home Area. The home area of an animal is a restricted stretch of familiar territory within which it carries on its normal activities during at least part of the year. The total range comprises all that territory in which it lives or through which it travels in its search for food, mates, or shelter. Sometimes the home area is coextensive with the total range. Frequently, though, it is not. Many amphibians live during the nonbreeding season in areas quite distinct from their breeding sites, and many snakes hibernate in communal dens which may be far from their normal activity ranges. The total range of an animal may include several home areas and the routes traveled between them.

The phrase "home range" appears frequently in the literature. It has been used both for total range and for home area, frequently with no attempt to distinguish between the two. This has led to much confusion, and makes it difficult to compare published results for different species. It seems better to drop "home range" and use the more precise terms "total range" and "home area."

The home area is not coextensive with a habitat, even in a limited sense. Thus a given woodland may be inhabited by a number of box turtles (*Terrapene*) but each turtle is restricted in its wandering to a particular part of that woodland. The mechanism by which the turtle is restrained from roaming at random until turned back by some physical difference in the environment is unknown. We can understand that a woodland species might hesitate to enter a sunny, open field, but why should it hesitate to enter another part of the same woods? Some recognition of familiar landmarks, hence some memory and learning ability, plus a preference for, or at least a tendency to turn toward, that which is familiar, seem necessarily involved but have not been experimentally demonstrated.

The few studies of amphibians indicate that possession of a home area is normal and probably widespread in the group. In a region of high population density, the average nonbreeding home area of a Spadefoot Toad (*Scaphiopus holbrooki*) covered about 120 square meters. The home areas of different individuals did not overlap, although they were sometimes contiguous. Home areas of Green Frogs (*Rana clamitans*) varied in size from 20 to 200 square meters, with an average for adults of 64.8 square meters.

A home area is not necessarily permanent. Adult *Scaphiopus* sometimes move from one burrow to another, although the overall home area remains essentially the same. But subadult Green Frogs tend to change the location of their home areas. A young frog usually establishes a small home area near shallow, densely vegetated water. When the frog reaches a length of about 60 to 75 mm., it moves to a new home area near deeper, more open water. Sometimes it vacillates for awhile between the two before finally settling down in the adult home area.

Many reptiles are known to have home areas. Those of the turtles and snakes are sometimes, though not always, surprisingly large, and may overlap extensively. Home areas of the aquatic turtles *Chrysemys* and *Pseudemys* may include parts of two or more bodies of water, between which the turtles make frequent overland journeys.

Lizard home areas are usually small. Those of adult males of the Five-lined Skink (*Eumeces fasciatus*) were found to average about 190 square meters, though some were considerably larger. As with the Green Frogs, young of this lizard apparently shift their home areas.

When a Syrian Fringe-toed Lizard (*Acanthodactylus tristrami*) was repeatedly frightened away from its burrow, it seemed to be thoroughly familiar with the terrain for about 40 square meters around the entrance to the burrow. If it was driven to the limits of this area, it became uncertain, stopped, and changed direction (see Fig. 9-1). This suggests, though it does not prove, that this animal had a very small home area.

Home Site. Not to be confused with the home area is the home site, a point within the home area to which the animal regularly returns when not foraging for food or carrying on other activities. It may be a burrow, a particular leaf on a tree (as for some tree frogs) or simply a restricted area, say a strip along the bank of a stream, within the home area.

Individuals of two species of salamanders, *Hemidactylium scutatum* and *Eurycea bislineata*, when kept in a terrarium, established home sites under bark fragments. Around its home site each developed a home area, through which it wandered freely while feeding. The terrarium was big enough to allow two animals to establish slightly overlapping home areas. When more than two were introduced, those that failed to establish home sites wandered at random through the areas of the others but did not feed and eventually starved to death.

Some frogs, such as the Squirrel Treefrog (*Hyla squirella*), do not seem to object to the proximity of others of their kind so that several may be found in one hiding place. Others (e.g., Spadefoot Toads, *Scaphiopus holbrooki*) are less socially inclined and do not share their home sites.

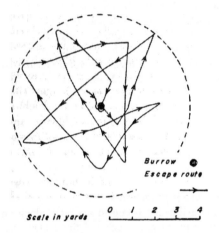

FIG. 9-1.
The route taken by a Syrian Fringe-toed Lizard, *Acanthodactylus tristrami,* when attempts were made to force it away from its burrow. [After Riney.]

Territoriality. When an animal defends the vicinity of its home site against intruders of the same species, it is said to show territoriality; the defended area is the territory. This type of behavior occurs most often among males, who frequently drive off other males during the breeding season, but sometimes females, and even young, may show territoriality.

The salamanders discussed in the preceding section warned intruders away from the immediate vicinity of their shelters. Frogs seldom show aggressiveness toward each other, but the male of the European Marsh Frog (*Rana ridibunda*) is said to chase other males away from its calling station. Territoriality is well developed in the South American frog, *Phyllobates trinitatus*. One frog will actually hop on top of another in an effort to drive it away. Adult females are the most aggressive, but even very young frogs may have territories. As with the salamanders, these are definitely feeding, not breeding, areas.

As yet, territoriality has not been adequately demonstrated in turtles or snakes. Turtles have been seen fighting one another in nature and so have snakes, but we do not know whether these battles represent true territoriality or are simply combat for mates.

Defense of territory is common in lizards. The Marine Iguanas (*Amblyrhynchus cristatus*) of the Galápagos Islands live along the coast, rarely going more than 12 to 15 meters inland. Along the shores, the young and females mass together, oftentimes piled on top of one another, two or three deep, all within about 9 meters of the water's edge. As many as 75 have been observed within a space of 9 square meters. The old males, which are much larger than the largest females, take positions between the massed females and young and the water. They are generally spaced from 1.5 to 3 meters apart, and each keeps to his own sunning territory. Every trespass by one male into the sunning terrain of another is the occasion for a fight. The contestants butt one another, each endeavoring to get his horny, knobby head beneath his opponent's chin. On some of the islands, the iguanas are found assorted in family groups, composed of a single large male with from two to four females. The families are separated from other groups by 18 to 45 meters of shoreline.

Among Black Iguanas (*Ctenosaura pectinata*) a social hierarchy is correlated with territoriality. Of nine males that occupied a stretch along an old cemetery wall, the dominant or tyrant male occupied the center position. He could and did invade the territories of all of the others. The one on his left and the one on his right were lower in the hierarchy. They could invade the territories of those to their left or right, respectively, but not the territory of the tyrant. The three on each end of this chain apparently had to tolerate the invasions of the tyrant or the subtyrants, but themselves could not invade any other territory.

Migration. The words "migration," "emigration," and "immigration" are sometimes used loosely, almost as if they were interchangeable. We here consider migration to mean a journey to *and return from* some place in correlation with a seasonal activity such as breeding. Emigration means a moving out from a region with no return implied. Immigration, moving into a region, is simply the obverse side of the coin since obviously, if an animal moves away from one place it must move into another.

True migration has been little studied in the amphibians. For most species we do not know whether the animals actually seek to return to the ponds from which they emerged or in which they had bred before. They may simply move to the nearest suitable body of water, or to the place where they hear other individuals of their own kind calling. Again, when the animals leave the ponds at the close of the breeding season, we do not know whether they return to the home areas they occupied before or whether they simply spread out until they find suitable territory in which to set up new nonbreeding home areas.

In England, the Natterjack (*Bufo calamita*) may use now one pond, now another. On the other hand, the Common Toad (*Bufo bufo*) is reported to resort to certain ponds year after year, while other ponds, just as suitable and accessible, are ignored. A colony of toads (probably *B. bufo*), under observation for ten years, migrated regularly between the breeding pond where they spent the summers and a sand pit in which they hibernated over the winters, a distance of 3 kilometers. Adult Green Frogs (*Rana clamitans*) are known to have returned to the home areas they occupied before they migrated to the breeding ponds. Individuals of the California Red-bellied Newt (*Taricha rivularis*) show a strong tendency to return year after year to the same breeding site. Of 262 males marked from a single pool in 1953, 85 percent were retaken in subsequent years, almost all from the same pool. (The newts spend the nonbreeding season on the mountain slopes, often far above the stream.)

Many snakes return year after year to the same hibernation dens. Desert Tortoises (*Gopherus agassizi*) have also been found to migrate, although over very short distances, between winter hibernation dens on the gravelly banks of washes and summer home areas on the benches above.

Reptiles and amphibians are obviously prevented by their limited powers of locomotion from such spectacular migrations as those of the birds. It seems probable, though, that the Green Turtles (*Chelonia mydas*) do migrate several hundred kilometers, from the Miskito Cays off the coast of Nicaragua, to lay on Tortuguero Beach, a strip about 30 km. long on the coast of Costa Rica. At least the big sea turtles disappear from the Nicaraguan fishing grounds in late May and early June, and soon after great flocks of turtles show up to lay their eggs and mate at the Costa Rican beach.

Homing. The behavior patterns so far discussed all involve some sort of homing, that is, the ability of an animal to find its way back to a particular spot. The simplest type of homing involves only the return of the animal to its home site from some point within its home area. That many herptiles possess this ability is a simple matter of observation. Indeed, the possession of a home site by an animal that normally wanders abroad in search of food or other necessities is in itself evidence that the animal has both the inclination and the ability to return to a particular spot.

A Gopher Tortoise (*Gopherus polyphemus*) was trapped in its burrow

and released in a nearby fire lane about 140 meters from the mouth of the
burrow (Fig. 9-2). "The time was 10:20 A.M. A few minutes later . . . it
had already moved to within 45 m. of the burrow. It progressed at a
leisurely but steady pace, pausing momentarily now and then to nip at
some vegetation. When nearly opposite the burrow site the turtle turned
at an angle and headed directly for the spot, which was about 18 m. away
from the fire lane and screened by brush at the edge of the fire lane. It
arrived back at the burrow at 10:32. I then took it about 180 m. E. of the
burrow into thick brush and released it. The time was about 10:40. I
smoothed over the entrance to the burrow after digging out the trap and
departed. When I checked at 2:30 P.M. I discovered tracks leading into the
burrow, probably of the individual handled in the morning." (From unpublished notes of J. N. Layne.)

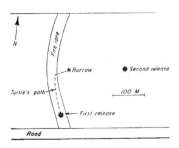

FIG. 9-2.
Route taken by a Gopher Tortoise,
Gopherus polyphemus, returning to
its burrow after release.

Off the coast of Nicaragua, the big, marine Green Turtles (*Chelonia mydas*) spend the day grazing in the shallow flats of turtle grass and return at night to rest, each on its own particular rock or coral head. For some this involves a daily round trip of 6.5 to 8 km.

Many frogs (e.g., species of *Leptodactylus, Hyla, Rana*, and *Bufo*) are known to have the ability to return to a chosen resting spot.

Even this simple type of homing raises some interesting psychological questions. Thus, Green Frogs (*Rana clamitans*) are notoriously slow at learning a maze. Yet in nature they show a strong tendency to return to a particular spot. It may be that the sensory cues presented to them in the maze are not of the same sort as the cues by which they guide themselves under natural conditions. Kinesthetic memory, that is, a tendency to repeat the muscular movements by which the animal achieved its goal (reached home) before, has been suggested. But it certainly is not the whole explanation, for the animal surely does not begin its return trip from the same place every time. There must be some recognition of familiar landmarks, hence some mechanism by which sensory impressions are stored within the brain (memory).

The homing involved in migration almost surely requires more than this. It is unlikely that the animal remembers landmarks along a route traveled once many months before. Nor, since most members of a migrating group arrive at the goal within a few days or weeks of each other, does it seem probable that the animals simply wander at random until they reach familiar territory. If the Green Turtles (*Chelonia mydas*) do indeed cross the open Caribbean to return to the nesting beaches, as the evidence indicates, they probably rely on celestial navigation as birds have been shown to do. Carr suggests the possibility "that green turtles use the heavens for guidance (as the turtle men say they do) and have the same spectacular guidance equipment (an inherent goal sense, a light-compass sense, a time correction sense, and a displacement sense or map sense) that other animals have been found to have."

To summarize briefly, seasonal migration is probably initiated by some physiological change in the animal, such as the release of hormones in the blood during the period of gonadal growth preceding the breeding season. These changes affect the nervous system so that the animal responds to stimuli ignored at other times, and begins to move in the direction indicated by the stimuli. For the amphibians, these stimuli seem to be relatively simple and direct—the calling of other frogs at the breeding pond, the odor of marsh vegetation, and the like. They may be nonspecific, so that the animal simply moves to the nearest suitable breeding site, or specific so that the animal returns to one particular place. The more highly evolved turtles are able to use more complex guidance mechanisms. Probably once the animal has reached familiar territory, it reverts to the sensory cues by which the home area was formerly recognized.

The possession by animals of such guidance equipment may help explain a third type of homing, the return of an animal to its home area when it has been removed to some place outside of its total range. This type of homing is difficult to demonstrate conclusively. Either the total range must first be worked out, or the place at which the animal is released must be so far away, or across such inhospitable terrain, that the point of release is surely beyond the limits of territory familiar to the animal. But if the distance is too great, or the intervening terrain too hostile, the animal may not be able to survive the journey back, even though it has both a desire to return home and some built-in mechanism that tells it which way to head.

It is sometimes possible to demonstrate, though, that an animal released in unfamiliar territory is able to orient itself and head in the right direction to get back home. When the California Newts (*Taricha rivularis*) referred to in the preceding section were displaced either upstream or downstream for distances of up to 0.64 kilometers, they moved in the proper direction to return to the home pool. Beyond this distance, they seemed lost and wandered at random. They did not use celestial navigation, for even when blinded they were able to orient successfully. Nor did they direct their movements by the slope of the ground. It is possible that the sensory basis of the guidance mechanism in these salamanders is a chemical one (odor) and the newts simply smelled their way back home.)   .

Box Turtles (*Terrapene c. carolina*) were carried from 0.45 to 9.0 km. from their home areas and released within an active military post surrounded by ditches, roads, fences, and buildings. It is highly probable that the release points were beyond the limits of territory familiar to the turtles. On bright, sunny days, many of them could and did orient themselves and begin traveling in the direction of their homes, but on cloudy days they seemed lost. When two turtles, one whose home was eastward and one whose home was westward, were released at the same time and place, the former headed east and the latter west. Several of the turtles, kept in a storage pit for some weeks before release, "homed" to the pit instead of to their former home areas.

Even though they do not migrate, these turtles were apparently capable of celestial navigation, guiding their movements by the position of the sun. Indeed they were frequently seen to stop walking and turn their heads as though sighting the sun. If turtles heading in the right direction were shielded from the sun but allowed to see a reflection of it in a mirror, they turned and moved toward the mirror. It would be interesting to see if turtles reared so that they never caught sight of the sun showed any ability to home.

As with the newts, the guidance mechanism of the turtles seemed to work only within restricted geographic limits. Of ten released more than 140 km. from home, seven headed in a direction that corresponded to the direction last chosen when close to home. This same tendency to head in the direction last chosen was also shown by a few turtles released less than a mile from home.

Emigration. As we said above, emigration means moving away from a place with no return implied. It is a one-way passage. Occasionally a drastic change in a habitat results in a mass movement of an entire population. These spectacular emigrations from drying ponds have been reported for frogs and turtles. A mass emigration may sometimes result in the occupation of a new habitat, an extension of range, or an increase in the rate of spread of a mutant gene through the species. Such events are rare and sporadic, though, and probably have much less effect on population structure, geographic distribution, and gene flow than does the inconspicuous but constant emigration of individuals.

Marking experiments on the Cliff Frog (*Syrrhophus marnocki*) showed that population density has an effect on emigration. When the population density of an area was lowered by removal of frogs, 19 out of approximately

42 frogs from a neighboring population moved into the area of low density. On the other hand, when 25 frogs were introduced into an area of high population density, none established residence and apparently one of the native frogs moved out. This suggests that intraspecific competition may be a potent cause of emigration.

This same study showed that ordinarily juveniles are the ones that tend to move out to establish new residences. Home areas of adults averaged 24 m. in diameter, but juveniles were recaptured at distances of 112 to 300 m. from the place of first capture. This tendency for the young to spread out from the area where they were born has been noted for many species.

Because of the difference in size, it is likely that the adults and young are not in direct competition. However, since in any population the number of young at the time of hatching (or of metamorphosis) is apt to be very large, severe overcrowding would result unless there was some mechanism for the dispersal of the juveniles.

It is probable that the tendency of the young to emigrate is not a direct response to overcrowding. The young may simply have a greater innate tendency to wander than the adults. Or more probably the young respond to stimuli to which the adults do not respond, or to which they respond in a different way. Young Green Frogs (*Rana clamitans*) emerging from a pond do move away from it, possibly as a negative response to the stimuli presented by a large, open body of water. Those that find suitable, smaller ponds, remain in them until the end of the following summer, when they have attained adult size. They then move to adult-type habitats in the vicinity of deeper, more open water. Since dispersal of the young does reduce overcrowding, it probably has positive selection value.

Obviously, hatchlings of aquatic and marine turtles do not simply wander at random, but orient themselves and head in the direction most likely to take them from the nest on shore to the water even though it is out of sight. A number of different stimuli have been suggested as providing the cues by which the hatchlings guide themselves; they may move toward a broad, open horizon, toward maximum areas of open, illuminated sky, toward light reflected from surf, in the direction of the downward slope of the ground, or away from a broken horizon or dark objects such as trees or bushes.

Hibernation and Aestivation

Hibernation and aestivation are responses to changes in climatic conditions and could equally well have been discussed in Chapter 8. We are including them here because we wish to stress some of the behavioral aspects of these responses.

Hibernation. Hibernation is an inherent, regular, and prolonged period of inactivity in the wintertime. It is characteristic of many herptiles of the temperate regions, though not, of course, of the tropics. With the approach of cold weather, the animals seek shelter under rocks or logs, in crevices, dens, or burrows in the earth, or under water, and pass into a torpid state. A decline in temperature is usually the trigger that sends the animals into hibernation. In mild autumns they may remain active much later than usual. There is evidence that for some forms (e.g., the frogs *Rana pipiens* and *Rana clamitans*), differences in temperature modify the stimulus-response mechanism. When the temperature is high, the animals tend to move toward the light, when the temperature drops, the animals move away from the light, and thus into sheltered situations where they are also protected from the cold. Once they have gone into hibernation, they do not ordinarily emerge until the following spring.

For some species, time of onset of hibernation seems to be innate. The animals withdraw within a week or two of the same time each year, regardless of weather conditions. Here the response of the animal may be modified by hormonal levels in the blood.

Hibernation may also involve physiological changes, such as a lowering of the metabolic rate and of bodily water content. Leopard Frogs (*Rana pipiens*), exposed to low temperatures in the summer, survived less than half as long as others exposed to the same temperatures in winter. Animals that hibernate may need the prolonged period of inactivity for the maturation of the sex cells. British Newts (*Triturus*), kept under constant temperature conditions in the laboratory, do not hibernate. Usually animals so kept fail to breed in the spring.

In general, adult herptiles go into hibernation before the young. Also, small species tend to emerge from hibernation earlier than large ones. In northeastern United States, the Northern Spring Peeper (*Hyla c. crucifer*) appears in the spring a month or two before the large Bullfrog (*Rana catesbeiana*). This is probably because a small animal has a greater surface area in relation to body volume than a large one and so can warm up its body more rapidly by absorbing environmental heat.

Hibernation does not seem to have reached the level of development among amphibians that it has in reptiles. It is not strongly fixed in many salamanders, since even in localities where most of them hibernate, occasional individuals may come up and wander around on favorable days. Most salamanders hibernate on land, under logs, in stumps, or in crevices in the ground. Some, such as the Tiger Salamander (*Ambystoma tigrinum*), hibernate in water at the bottom of ponds.

Salamanders often suffer great losses in number because of their inability to pick suitable hibernacula (sing. hibernaculum, a place of hibernation). European Smooth Newts (*Triturus vulgaris*) are unable to dig but creep into cracks in the soil and worm their way down. They do not pick soil that is too moist or too dry, but they sometimes fail to go deep enough, or sometimes simply hide under boards where they do not have sufficient protection. Nearly 50 percent have been reported killed at one hibernating site.

Frogs resemble salamanders in hibernation pattern. Some, such as the Leopard Frog (*Rana pipiens*), dig down in the mud on the bottom of a pond, others, as the American Toad (*Bufo americanus*), burrow into suitable soil, still others may shelter under logs, in dead stumps, or in depressions below the leaf mold.

Amphibians frequently hibernate alone, but sometimes large numbers of frogs or salamanders are found at the same site. Usually though, the animals seem to have come together because only a few spots in the region offer the right combination of soil, moisture, and protection. They do not "ball up" the way snakes do in a hibernating den. Each one is isolated in the midst of the crowd and there is little or no social significance to the aggregation. One unusually large concentration of hibernating individuals has been reported for the Common Frog of Europe (*Rana t. temporaria*). The frogs were lying on the bottom (60 cm. deep) of a clear, slowly flowing stream. They were gathered into three groups, two containing between 100 and 200 individuals, and the other between 400 and 500. These gatherings may have been truly social in that the animals were attracted by the presence of others of their kind instead of by some feature in the physical environment.

Many reptiles do not differ markedly from the amphibians in hibernation patterns, but others, particularly snakes, show a very strong tendency to form denning aggregations. The same snakes will return to a den year after year, sometimes from distant points. Solitary hibernators are occasionally found, even in species in which the tendency to aggregate is strong, but still this social denning is the most characteristic feature of snake hibernation. Sometimes the animals coil around each other to form large "balls of snakes." Such balls possibly help conserve heat and reduce moisture loss and so are beneficial to the individual. Suitable dens may be used by several different species at the same time, and occasionally other herptiles are found hibernating with the snakes. A den excavated in England contained 40 adders, 10 toads, and many lizards.

A large den in Tooele County, Utah, has been studied for many years. Apparently bygone springs washed loose material from among·the stones and boulders on the shoreline of ancient Lake Bonneville, leaving large underground channels into which the snakes descend for hibernation. From this single den, 930 Great Basin Rattlesnakes (*Crotalus viridis lutosus*), 632 Striped Whipsnakes (*Masticophis t. taeniatus*), 127 Western Yellow-bellied Racers (*Coluber constrictor mormon*), 36 Great Basin Gopher Snakes (*Pituophis catenifer deserticola*), 2 Desert Night Snakes (*Hypsiglena torquata deserticola*), 2 Regal Ringneck Snakes (*Diadophis r. regalis*), and 1 Longnosed Snake (*Rhinocheilus lecontei*) have been recorded. There are 1,080 recapture records for rattlesnakes, 440 for whipsnakes, 28 for racers, 4 for gopher snakes, making a total population of 3,282 snakes for the den over a 10-year period. Obviously, not all of these snakes lived in the vicinity of the den during the summer months, and some must have traveled long distances to reach it in the fall.

How the snakes find their way to such a den is not known. It has been suggested that it may be partly a learned behavior, with the young following the scent trails of the returning adults and so becoming familiar with the route. But what stimulates the adults to return and what sensory cues they use to guide themselves remain unknown.

Retraherence. A temporary retreat from adverse weather conditions is often alluded to as hibernation. We prefer to use the term "retraherence" (from Latin *retrahere* = to withdraw) for this behavior to distinguish it from the rhythmic hibernation defined above. In northern Florida, the lizards *Eumeces laticeps* and *Cnemidophorus sexlineatus* have definite hibernation periods that last throughout the winter regardless of the weather conditions on any particular day. In the same region, two other lizards, *Sceloporus undulatus* and *Anolis carolinensis*, show retraherence. On cold days they stay in their retreats, but on warm, sunny days, even in midwinter, they come out to feed and bask in the sun. Hibernation and retraherence grade into one another and a species that normally hibernates in one part of its range may not do so in another part. The western Red-legged Frog (*Rana aurora*) definitely hibernates in the northern part of its range and at high altitudes in the southern part, whereas at low altitudes in the south it simply retreats from unpleasant weather, to emerge again as soon as conditions improve.

Aestivation. Aestivation, a regular, obligatory retreat from hot, dry climatic conditions, is an unusual phenomenon among herptiles. Many of them do show retraherence from conditions of drought and high temperature. South African Clawed Frogs (*Xenopus laevis*) bury themselves in the mud when the ponds in which they live begin to dry up, and many salamanders retreat into caves during hot, summer weather. Usually, though, such withdrawals do not seem to represent an innate seasonal rhythm, for a temporary amelioration of the weather is enough to bring the animals forth again.

The Sardinian Cave Salamander (*Hydromantes genei*) has been reported to aestivate in summer even in the laboratory where humid conditions were presumably maintained. Captive Australian Long-necked Turtles (*Chelodina longicollis*) tucked themselves away for weeks without feeding in the summer even though their ponds were kept filled with water. These may be examples of true aestivation, but whether they are accompanied by physiological changes similar to those reported for hibernating herptiles has not been determined.

Some herptiles (lizards such as *Heloderma*, *Tiliqua*, and many gekkonids, and the salamander *Plethodon richmondi*) accumulate stores of fat in their tails before the onset of unfavorable weather conditions, but this fat storage apparently is not accompanied by obligatory aestivation.

Feeding

The kinds of food eaten by herptiles were discussed in Chapter 8. Here we are more concerned with the means by which the animals recognize and capture their food.

Adult frogs and toads are strictly carnivorous. Among the terrestrial forms, the sight of a moving object of the appropriate size is the stimulus that brings forth the snapping response by which the food is taken into the mouth. A quiescent object, however edible, is completely ignored. The tendency to snap at any small, moving object sometimes gets an amphibian into trouble. On the Hawaiian Islands, the introduced Marine Toads (*Bufo marinus*) are attracted by the falling blossoms of the Strychnine Tree and catch them as they flutter to the ground. A toad with its stomach full of these flowers soon dies of strychnine poisoning.

It is possible, though, for a frog to learn to recognize unpalatable objects and to suppress the snapping response. Offered a hairy caterpillar, a frog snapped it up and then promptly got rid of the mouthful. After the caterpillar had been presented several times, the frog hopped up close to it, slowly jutted its head forward, and touched the caterpillar with its extended tongue. Then it quickly withdrew its tongue and thereafter refused to respond to the caterpillar. Common toads of Europe (*Bufo bufo*) learn not to snap at bees after they have been stung several times. Other toads of the same species, though, discover how to capture the bees without being stung and may be found at night waiting at the entrance of a hive to feed on the latecomers as they arrive home.

A toad (*Bufo bufo*) kept as a pet by Boulenger, would swallow so many earthworms in succession that after a time they were passed alive. Apparently, in this individual at least, the snapping reflex was not inhibited by a sense of repletion.

Some tadpoles are entirely herbivorous, foraging for algae in the water or scrapping them off rocks on the bottom with their horny jaws. Obviously, the stimulus-response mechanism involved in this type of feeding is quite different from that shown by the carnivorous adult. Adults and young respond to different stimuli by different muscular movements to accomplish the same end—getting food. This means that metamorphosis must include a drastic shift in the neurological basis of feeding behavior.

Both adult and larval salamanders are carnivorous. They are more inclined to take quiescent food, such as the eggs of fish or amphibians, than are the frogs. Obviously they must depend on other sensory cues than the sight of a moving object. Blinded *Ambystoma* larvae respond to the odor of food. Some newts (*Triturus*) will feed after both optic and olfactory nerves have been cut, suggesting that sensory receptors in the skin may also function in the perception of food.

Diurnal lizards that feed largely on insects resemble the frogs in their tendency to snap at any small, moving object. Herbivorous lizards probably also rely largely on sight, though odor may play a part in the recognition of suitable food items. Snakes may use either sight or smell in finding their prey. Chemical stimuli seem to be most important to the burrowing forms and also to the aquatic snakes and turtles. The specialized heat receptor organs of the pit vipers probably function primarily in guiding the direction of the strike once the prey has been located. Snakes that eat eggs are sometimes fooled by glass nest eggs or smooth, round stones and swallow these indigestible objects, suggesting that they rely largely on sight rather than smell in identifying the eggs.

Some herptiles actively forage for their food, others find a favorable location and wait for the food to come to them. Nocturnal insect-eaters like the tree frogs and geckos are attracted to lighted windows as are the insects on which they feed. Green Frogs (*Rana clamitans*), which eat recently transformed young of their own or other species, gather at the water's edge when the young are emerging. This may be an important selective factor impelling the young to move away from the breeding ponds.

A few species that habitually lie in wait for their prey have developed lures to draw their victims within reach. Best known is the Alligator Snapping Turtle (*Macroclemys*) which has a pink, wormlike projection on its tongue. The hungry turtle lies quietly on the bottom with its mouth open and wiggles the lure. A fish that swims up to investigate is easily caught. Juvenile snakes of some species of the families Boidae and Viperidae have the tip of the tail modified. It is bright yellow in color and wormlike in appearance, and has been reported to serve as a lure. In the presence of potential prey, the snake lies quietly with the tail raised and writhing like a worm or an insect larva. Both frogs and lizards may be attracted by these caudal lures.

Finally, it should be mentioned that poison among the snakes is largely associated with food gathering. The rattlesnakes capture their prey by striking and injecting venom. If the animal is large, the snake may then draw back to wait until the poison has taken effect; if it is small, the snake will retain its hold until the victim ceases to struggle and can be swallowed. On the other hand, the poisonous lizards (*Heloderma*) feed mostly on eggs, fledgling birds, and nestling mammals, which can be captured without the aid of venom. Their poison may be used to reduce struggles in prey and probably originated as a food-gathering mechanism in some lizard ancestor of the past.

Defense

Like most animals, amphibians and reptiles avoid trouble and are seldom aggressive. Fighting or threatening behavior usually takes place between members of the same species in defense of territory or in combat for a mate. At the approach of a potential predator, a herptile seeks to avoid notice by remaining perfectly still or by slipping quietly away. A nearer approach may result in more active escape measures. True frogs (*Rana*) jump for the water, toads (*Bufo*) and Ground Skinks (*Lygosoma*) retreat into their burrows or under sheltering objects, and Race Runners (*Cnemidophorus*) count on their speed to outdistance an enemy.

When actually cornered, amphibians and reptiles are capable of a rather surprising variety of defensive maneuvers. A frog may release water stored in its cloaca, thus reducing its weight for jumping and perhaps distracting its pursuer. Some salamanders and lizards are able to break off a part of the tail when hard pressed. The detached portion wiggles violently, thereby distracting the predator, while the main body of the animal escapes unnoticed. Frequently the tails of these species are brightly colored. Toads inflate their lungs and assume a characteristic defense posture with the head lowered, eyes closed, hind legs spread, body tilted toward the aggressor. This results in both an apparent and actual increase in size which may effectively discourage small predators, especially snakes which do not generally tackle objects too large for them to swallow. Many other frogs also inflate and assume defensive poses when threatened. Salamanders, too, may have defense postures. *Ensatina eschscholtzi* stands high on its legs when annoyed and waves its tail, which bears poison secreting glands, at its attacker.

Some snakes roll up into balls and tuck their heads under their coils when threatened. Sometimes the tail is elevated and presented to the attacker rather than the more vulnerable head. A threatened Rattlesnake (*Crotalus*) usually coils in the striking position, but when attacked by an ophiophagous Kingsnake (*Lampropeltis*) it reacts in a very different way. It inflates its body, lowers its head to the ground, arches the middle part of its trunk, and seeks to fend off its attacker by blows with the arched body loop. A Rattlesnake placed in a jar from which a Kingsnake has been removed assumes

FIG. 9-3. The defense posture of the Common Toad of England, *Bufo bufo*. Based on a specimen posed after death. [Photograph by W. S. Pitt.]

the characteristic "kingsnake defense posture," showing that it is the odor rather than the sight of its enemy that evokes the response.

One of the most spectacular defense displays is that of the Hognosed Snake (*Heterodon*). It rears up, flattens and spreads its neck, hisses, and may even strike at the intruder, though always with its mouth closed. If this threatening display does not drive off the enemy, it rolls over on its back, lolls out its tongue, goes limp, and plays dead. This death feigning has been considered the same as the hypnotic state of tonic immobility which can be easily induced in a number of herptiles. But the fact that the "dead" *Heterodon*, when placed on its venter, promptly flops over on its back again indicates that more is involved.

Mucous secreted by the skin glands of a threatened amphibian make it slippery and difficult to hold. Many frogs and salamanders also secrete poisons that make them either distasteful or actually dangerous food for a predator.

As a last resort, some salamanders and frogs will bite and a large Congo Eel (*Amphiuma*) can inflict a painful wound. The effectiveness of the biting response of the larger reptiles, and particularly of the poisonous snakes, is too well known to need stressing here.

The descriptions above include only a few of the behavioral responses of which herptiles are capable. Some others were discussed in the chapters on reproduction. Most have not been analyzed, even to the extent of showing to what stimulus the animals are responding, and the whole field is one in which there is still much work to be done.

Collateral Reading and General Reference

Angel, F. *Vie et Moeurs des Amphibiens*. Paris: Payot, 1947.
    Vie et Moeurs des Serpentes. Paris: Payot, 1950.

Hebb, D. C. *The Organization of Behavior*. New York: Wiley, 1949. (Not concerned with the herptiles, but an excellent account of recent work on the neurological basis of behavior.)

Maier, N. R. F. and T. C. Schneirla. *Principles of Animal Psychology*. New York: McGraw-Hill, 1935. (Contains sections on the psychological basis of behavior in amphibians and reptiles.)

Noble, G. K. *Biology of the Amphibia*. New York: McGraw-Hill, 1931.

Oliver, J. A. *North American Amphibians and Reptiles*. Princeton: Van Nostrand, 1955.

Smith, M. *British Amphibians and Reptiles*. London: Collins, 1951.

MECHANISMS OF SPECIATION

THE EARLY, CLASSICAL STUDIES of evolution were concerned with the vast sweep of change through ages of time. At the lowest level, they dealt with the shift from species to species, genus to genus, over perhaps a million years, and paid little attention to the primary sources of the variations which are the building blocks of evolution. On the other hand, early genetic studies stressed the individual and the changes in its genes and chromosomes that made it different from other individuals of the same kind. Many geneticists were little concerned with fitting these changes into the broad canvas painted by the evolutionists, and there was little meeting of minds between the two schools. Now they are drawing closer together. The modern field of population genetics is concerned with the ways in which mutations are spread through animal populations to make them different from other populations of the same species, while the accumulation of large series of some fossil forms has made possible the application of genetic principles to at least a few extinct populations. To understand evolution we must study both the broad outlines of long-term changes in form and the minute details of changes within the nucleus of the cell.

SOURCES OF VARIATION

The basic sources of variation are the changes (mutations) in the genes and chromosomes.

170

Evolution of Karyotypes

Recently, great strides have been made in the study of the chromosomes of amphibians and reptiles. The karyotypes (the number, size and shape of the chromosomes) of many species belonging to various families have been described in detail and from these descriptions a pattern of the evolution of the karyotypes is emerging. Two generalizations are possible: first, that the primitive ancestral stocks had chromosomes of two basic types, large (macro) and small (micro); second, that the primitive condition was to have acrocentric chromosomes, that is, chromosomes in which the centromere, the point to which the spindle fiber attaches during cell division, is at one end.

Fusion of Chromosomes. Evolution in both the reptiles and the amphibians seems to have been along the line of fusion of acrocentrics in the region of the centromeres so that, in general, the more modern, specialized forms have smaller numbers of chromosomes but tend to have more V-shaped, metacentric ones, in which the centromere is at the point of the V, and fewer I-shaped acrocentric ones. Furthermore, it appears that in the main macro-acrocentrics have fused to form macro-metacentrics and that micro-acrocentrics have fused to form micro-metacentrics rather than that micro- and macro-acrocentrics have fused.

Curiously, though, the amphibians seem to have undergone more fusion than have the reptiles. Only among the more primitive anuran families—Leiopelmidae, Discoglossidae, and Pipidae—do we find acrocentrics present at all. Fusion of acrocentrics to form metacentrics has not proceeded apace in the reptiles; only the lizard *Chamaeleo* is known to lack acrocentrics. The state of fusion is also more subject to individual variation in the reptiles. Thus some Oregon Alligator Lizards, *Gerrhonotus multicarinatus scincicaudus,* have eighteen acrocentrics and two metacentrics, but others of the same race have twenty acrocentrics and only one metacentric.

The salamanders differ from the frogs and reptiles in karyotype pattern. The number of chromosome pieces is greater in the more primitive salamanders and the chromosomes are not so readily divisible into macro- and micro-chromosomes as are those of the Anura and Reptilia. Although the same method of chromosome evolution by centric fusion seems to have occurred in this group as in the others, additional mechanisms must also have been at work. It has been suggested that a series of losses and fusions of small chromosome fragments has played a part in the evolution of salamanders as compared to the simple fusion of acrocentrics into metacentrics in the other herptiles. There seems to be no close cytological connection between the Caudata and the Anura.

The hypothetical evolution of karyotypes is illustrated by the diagram (Fig. 10-1) which shows how the fusion of parts decreases the number of both micro- and macro-chromosomes but increases the number of meta- centrics as compared to the number of acrocentrics.

Number of Chromosome Arms. The basic number of chromosome arms, the "Nombre fondamental" (usually abbreviated N.F.), is remarkably con- stant for the various groups of amphibians and reptiles. The N.F. of any stock is the number of all the chromosome arms, in- cluding the arms of both micro- and macro-chromo- somes. To calculate it one simply adds the number of acrocentrics to twice the number of metacentrics. Some related forms that are quite different in chromosome numbers have the same N.F. For exam- ple, the only two Apoda that have been examined cy- tologically, *Ichthyophis glutinosus* and *Uraeotyphlus narayani*, have respectively 42 and 36 chromosomes, but the former has 32 I-shaped acrocentric ones and 10 V-shaped metacentric ones, giving it an N.F. of 52, whereas the latter has 20 I-shaped acrocentrics and 16 V-shaped metacentrics so that it likewise has an N.F. of 52.

FIG. 10-1.
The hypothetical evolution of anuran karyo- types. 1. Hypothetical stem form. 2a. *As- caphus*, with 12 metacentrics and 30 acro- centrics. 2b. The situation in *Alytes* and *Xenopus*, with 12 metacentrics and 24 acro- centrics. 3. *Discoglossus*, with 20 metacen- trics and 8 acrocentrics. 4. The situation in most higher forms which have only meta- centrics, based on *Bombina*, with 24 meta- centrics. Most modern forms have 22 to 26 metacentrics. [After Wickbom.]

Little has been accomplished in correlating shifts in karyotypes with precise evolutionary modifications of structure. Nevertheless, these studies strikingly confirm some of the conclusions derived from paleontological and anatomical studies. They suggest the derivation of the anurans and the reptiles from a common stem and the wide separation of the Caudata. They confirm the primitive position assigned to the frogs of the families Leiopelmidae, Discoglossidae, and Pipidae. And they offer much hope for the future clarification of some perplexing problems of classification.

Appendix B gives a summary of the chromosome numbers of those amphibians and reptiles that have been adequately studied cytologically.

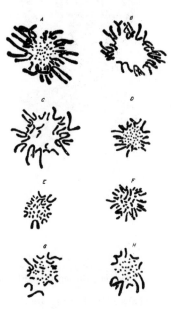

FIG. 10-2.
Diploid chromosomes of several genera of amphibians and reptiles. A. *Cryptobranchus*. B. *Triturus*. C. *Rhacophorus*. D. *Clemmys*. E. *Alligator*. F. *Gekko*. G. *Elaphe*. H. *Naja*.

Genetics

Herptiles are relatively slow breeders. Many of them take two or more years to reach maturity. In addition, the offspring of the amphibians usually pass through an aquatic larval stage before they metamorphose to a point where their characters can be compared to those of their parents. Also, it is usually more difficult and expensive to raise large numbers of herptiles in a laboratory or under controlled conditions than it is to raise fruit flies or mice. For these reasons, detailed genetic studies of amphibians have lagged, but a start has been made. Most of the work so far has been concerned with the inheritance of color pattern.

Genetic Studies on Amphibians. The little Greenhouse Frog of the West Indies (*Eleutherodactylus ricordi planirostris*), which has become so widespread in Florida, has two distinct dorsal patterns (Fig. 10-3) called striped and mottled. When two striped individuals are mated, their offspring are either all striped or some are striped and some mottled, and the same is true of striped × mottled crosses, but when two mottled frogs are mated,

FIG. 10-3. Striped (left) and mottled (right) patterns of the Greenhouse Frog, *Eleutherodactylus ricordi planirostris*. [From Goin, *Studies on Eleutherodactylus*, University of Florida Press, 1947, by permission.]

only mottled young are produced. Apparently then the gene producing the striped condition (S) is dominant and the gene for the mottled condition (s) is recessive. A random sample taken to determine the phenotypic ratio of the population at Gainesville, Florida, indicated that .2443 of the population were mottled and thus presumably homozygous recessives. Once the number of homozygous recessive animals in a population is known, it is possible, by means of the Hardy-Weinberg equilibrium formula, to work out the relative frequency of the two genes in the population. The equilibrium formula is:

$$q^2 = SS$$
$$2q(1 - q) = Ss$$
$$(1 - q)^2 = ss$$

Since we know that *ss* = .2443, we can solve the last equation to find the value of *q* and then substitute this value in the other equations. Thus it is possible to figure out the probable gene distribution in the Gainesville population, as shown below.

$$(1 - q)^2 = .2443 \qquad ss = 24.43\% \quad \text{Mottled}$$
$$2q(1 - q) = .4999 \qquad Ss = 49.99\% \left.\vphantom{\begin{matrix}a\\b\end{matrix}}\right\}\text{Striped}$$
$$q^2 = .2557 \qquad SS = 25.57\%$$

Also, if the frogs mate at random, so that a striped frog is as apt to mate with a mottled one as with another striped one, then striped and mottled patterns should appear in the offspring in about the same proportions as in the parent population. Of 1,395 frogs hatched from eggs collected in the same area, 354 were mottled and 1,041 were striped. Theoretically, of these 1,395 offspring, 24.43 percent or 341 should have been mottled and 75.57 percent or 1,054 striped. This close correspondence between the theoretical and actual patterns in the offspring is in agreement with the view that mating is random.

Similar studies on other species of *Eleutherodactylus* indicate that the striped pattern is also inherited as a dominant in such forms as *E. alticola*, *E. nubicola*, *E. gossei*, *E. pantoni*, and *E. nasutus*. It seems that this pattern, so widespread in the genus, is due simply to homologous genes in many of the species. Other patterns found in frogs of this genus, such as the presence of a narrow, middorsal stripe, or of a shield-shaped, light area on the back (called picket) apparently are likewise expressions of simple color pattern genes.

The uniform dorsal coloration in the frog described as *Rana burnsi* is due simply to a single dominant gene that is allelomorphic to the recessive gene causing the spotted pattern in the Leopard Frog, *Rana pipiens*. Not only this, but there is strong evidence that the dorsal spotted patterns of two quite distinct species, *Rana areolata* and *Rana palustris*, are expressions of the same gene and that these three species have this corresponding gene locus in common. In *R. areolata*, *R. palustris*, and the spotted "wild type" of *R. pipiens*, the genotype is pure recessive (*bb*) but in the *R. burnsi* form of *R. pipiens* it is *Bb*. The form *R. burnsi* is rather rare. It is restricted geographically to northern Iowa, southern Minnesota, and northwestern Illinois and constitutes less than 5 percent, often less than 1 percent, of the total *R. pipiens* populations in the areas in which it occurs. The absence of any *BB R. burnsi* frogs among those tested in breeding experiments apparently results simply from the rarity of the gene in the population. Parenthetically, it might be stated that the form *R. burnsi* obviously should not be considered a separate species.

Sometimes the inheritance of color pattern has a more complex basis. The European Fire Salamander, *Salamandra salamandra*, has both a striped phase and a melanistic phase which is associated with stunted growth. The melanism varies greatly in intensity in different individuals and is apparently controlled by more than a single pair of allelomorphic genes.

Genetic Studies on Reptiles. In reptiles as in amphibians the few cases of inheritance that have been analyzed genetically deal with color pattern. The population of garter snakes (*Thamnophis s. sirtalis*) around Lake Erie includes a number of black individuals as well as ones with the normal striped pattern. In some places the black ones make up as much as a third of the population. It has been shown rather conclusively that the melanistic pattern is due to a recessive gene "*b*," so that garter snakes of the genotypes *BB* or *Bb* have the normal striped pattern but those of the genotype *bb* are black.

A more striking instance among snakes, albeit admittedly less well understood, is that of the Kingsnakes of California, in which a striped form (originally called *Lampropeltis californiae*) breeds with a ringed form (*Lampropeltis getulus boylii*). Enough data have been accumulated to demonstrate that both striped and ringed offspring are produced by mothers of both types, thus indicating a mendelian basis for the patterns. The evidence is too scanty as yet to determine which of the two patterns is dominant and whether or not there is random mating in the population. Finally, some of the offspring have a somewhat aberrant pattern that is obviously allied to, although different from, the striped pattern. Whether this is due to a modifying gene of some sort is not known.

SPECIATION

The appearance and spread of mutations through a population is not in itself enough to cause speciation. Dr. Ernst Mayr has stated that "the problem of the origin of the species is a problem of the origin of discontinuities, for although evolutionary change alone may lead to modifications of previously existing species it cannot lead to their multiplication." This is an excellent summary of the situation. If a discontinuity develops between two parts of a population and this discontinuity is maintained long enough to allow for genetic reconstruction and the development of isolating mechanisms, speciation has occurred.

Therefore, the first step in the study of speciation is the search for the causes of discontinuities. They may arise either:

1. Instantaneously, by means of polyploidy. This has not been shown to be a cause of speciation among herptiles.

    2. Gradually, by:

        a. Geographic separation, in which the populations occupy different areas and are separated either by distance or by a physical barrier such as a mountain range.

   b. Ecologic separation in which the populations occupy essentially the same area but different habitats within the area.

Geographic Discontinuities. A population may have a wide geographic range. Since environmental conditions differ from one part to another of the range, the members of the population are exposed to different selective forces in different areas. When the change in conditions is gradual, as the increase in temperature as one moves from north to south in the northern hemisphere, characters in the population may also show a gradual change. For example, the average number of ventral plates of the Eastern Kingsnake (*Lampropeltis g. getulus*) increases from north to south. In southeastern Virginia it is 211, in North Carolina, 212, in South Carolina, 213, in southeastern Georgia, 216, and in northern Florida 218. Such a character gradient is called a cline. The Kingsnakes show similar clines in other characters, so that the part of the population at one end of the range differs noticeably from that at the other end.

In many species, characters do not show the smooth curve of clinal change. They are relatively constant over a part of the range, then shift rather abruptly. This may be because the environmental factors also change abruptly, as in the transition from woodland to open grassland, or because some partial barrier reduces the flow of genes from one segment of the population to another. (Differences in environment may themselves act as partial barriers by limiting the amount of contact, and hence the degree of interbreeding between the parts of the population.) When the characters by which one segment of a population differs from another change abruptly rather than gradually, the two may be named as subspecies or geographic races.

Ecologic Discontinuities. Geographic and ecologic discontinuities really differ only in degree, since ecologic discontinuities are microgeographic as well. Thus on Jamaica frogs of the genus *Eleutherodactylus* have certainly undergone ecologic speciation, but this speciation is at the same time associated with minor geographic differences. *Eleutherodactylus gossei* is widespread in the lowlands and on the dry lower slopes of the mountains. In the very humid cloud forest, which occurs at elevations of about 1,200 m., a larger, dark-bellied form, *E. nubicola,* is present. Above this, ranging into the wind-scrub on the top of Blue Mountain Peak, occurs *E. alticola,* smaller than *E. nubicola* and much less variable in color pattern than the other two. Obviously, the cloud forest is not only ecologically much different from the dry lower slopes but is also above them in a geographic sense. It seems useful, though, to maintain the term ecologic (sympatric) speciation to differentiate cases of this sort from the speciation resulting from discontinuities produced by simple geographic distance (allopatric speciation)· It is also true that, while all ecologic speciation is microgeographic, geographic speciation is also ecologic since pure geographic distance seems invariably to involve some change of ecologic conditions.

Genetic Reconstruction

Once discontinuities have developed in a population, two things in addition to mutation can bring about genetic reconstruction of the isolated forms. One of these is natural selection. Since no two areas are ever exactly alike, the two parts of the population are exposed to different selective forces and hence come to differ from each other.

The second factor leading to genetic reconstruction is random fixation. The individuals of a population vary. Many genes are present in all members, but other genes are present in some individuals and not in others. Thus some of the *Eleutherodactylus* discussed above have the gene for the striped condition and some do not. The sum total of the genes present in a population is known as the gene pool. When a segment of a population is separated from the rest, mere chance determines how its gene pool will differ from that of the parent population in relative proportions of genes. A gene rare in the one may be relatively common in the other, or it may be absent entirely. If one or both of the populations is small, the gene pools will come to differ further from each other through the process known as genetic drift. If only a few members of the population carry a given gene, and if they, by chance, fail to reproduce successfully, the gene will be eliminated from the population. If the population is very small, even relatively common genes may be eliminated. Conversely, a gene common in the population may come to be fixed, that is, to be present in all members of the population. In very large populations, in the absence of selection, the relative proportions of genes in the gene pool remain about constant.

There have been many excellent studies of geographic variation in the herptiles. We can determine from them some of the characters by which the populations differ, but too often we can do no more than guess as to whether these characters were fixed by selection or by random genetic drift within the population. Studies on the genetic differences between related species, and of the selective values of the different genotypes under different environmental conditions, remain an acute need.

Isolating Mechanisms

Differences of many kinds may develop between isolated populations, but only those differences that bring about genetic isolation lead to speciation. These we call isolating mechanisms. Once such mechanisms have developed, the formerly isolated populations may come to occupy the same area through a breakdown of barriers, but they will remain differentiated. Isolating mechanisms may be of various kinds. Some of them are outlined below.

I. Factors that inhibit crossing

    A. Breeding behavior differences

        1. In breeding sites
        2. In time of breeding
        3. In courtship patterns

    B. Physical differences

        1. In recognition characters
        2. In form of genitalia and related structures

II. Development of genetic incompatability

    A. Hybrid sterility
    B. Hybrid inviability
    C. Primary sterility

Different Breeding Sites. Good examples of differences in choice of breeding sites have as their *sine qua non* detailed life history studies, of which, unfortunately, there have been all too few. Still, species pairs are known in which the adults live together during most of the year, but move to different sites in the breeding season. Of the two toads found in England, the Common Toad (*Bufo bufo*) breeds in moderately deep water in ponds and canals, whereas the Natterjack (*Bufo calamita*) breeds in shallow water, often in small puddles only an inch or two deep.

Different Breeding Seasons. At Bloomington, Indiana, the American Toad (*Bufo americanus*) begins to breed during the last week in March, but Fowler's Toad (*Bufo woodhousei fowleri*) does not start until the middle or end of April. The two forms also differ in their calls and, to a large extent, in their choice of breeding sites. There is some overlap of the breeding seasons and during this brief period the two do interbreed to some degree. Furthermore, of the toads that breed in this period

of overlap, a large percentage of the individuals are more or less intermediate morphologically between the two species and their calls are apt
to be somewhat intermediate. None of the crosses discovered during the
period of overlap involved individuals near the mean of variation of either
of the respective species.

Isolation between the two forms is thus not complete. It has been suggested that they may have come together in relatively recent times, perhaps
because ecologic conditions in the region have been altered by human activities. Unless there is some selective pressure against the hybrids, the two
species may eventually merge.

It may well be that two of the Hognose Snakes, *Heterodon platyrhinos*
and *Heterodon simus*, are separated in part by a mechanism of this sort.
H. platyrhinos emerges fairly early in the spring whereas *H. simus* is definitely a hot weather snake. Perhaps *H. platyrhinos* copulates earlier in the
year than *H. simus*, but the actual time when *H. simus* copulates is still
unknown. Studies of this sort are much needed.

Different Courtship Patterns. Male turtles of the species *Pseudemys
scripta* go through a remarkably different liebespiel from that of the males
of *P. floridana*. In the former the male swims backward facing the female,
stroking her face and chucking her under the chin with the long nails of
his front feet. In the latter species both face in the same direction and the
male swims just above the female, his head bent downward close to hers.
He then turns his tail down and under the rear margin of her shell; at
this point the liebespiel ends and the two together sink slowly to the bottom, where presumably copulation takes place.

Different Recognition Characters. One has only to listen to a few frog
choruses to realize that related species may breed in the same pond but
have quite different voices. It is true that much yet remains to be found
out about the breeding voices of frogs. We can be fairly sure, though,
that they do function as isolating mechanisms because usually a female
Rana pipiens moves to and goes into amplexus with a calling *Rana pipiens*
rather than with the male of some other species. Undoubtedly the many
distinctive frog voices have a profound effect in keeping the forms isolated.

That size likewise may be effective is indicated in part by the geographic
distribution of many forms. The salamander genus *Plethodon* of the eastern
United States includes many species pairs that live and breed in the same
woods but are different in size. Since they do not depend on voice for recognition, and since they are usually found under rocks, logs, or leaf litter
where sight would be of little use, it may well be that the difference in
size is an important isolating mechanism. Detailed life history studies will

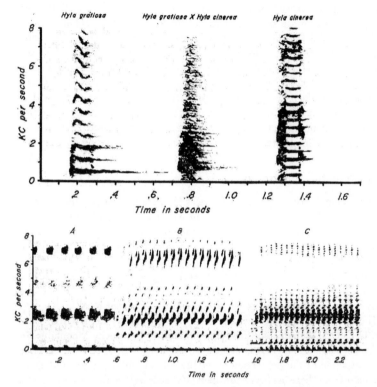

FIG. 10-4. Sound spectrograms of the voices of certain tree frogs of the genus *Hyla*. It seems probable that differences in voice such as these serve as isolating mechanisms in frogs. (Upper) Mating calls of *Hyla gratiosa* (left); hybrid, *Hyla gratiosa* × *Hyla cinerea* (center); and *Hyla cinerea* (right). All recorded near Archbold Biological Station, Highlands County, Florida. (Lower) Mating calls of three hylids. A. *Hyla femoralis*, portion of call recorded at Orange Springs, Florida; B. *Hyla versicolor*, entire call recorded at Hazelhurst, Wisconsin; and C. *Hyla arenicolor*, entire call recorded at Cave Creek, Arizona. [Spectrograms from Bogert, in *Animal Sounds and Communications*, A. I. B. S., 1961, by permission.]

probably some day give us a long list of isolating mechanisms based on recognition characters.

Different Genital Structures. Two Asiatic species of Lancehead Snakes of the genus *Trimeresurus* are strikingly similar both in color pattern and in external morphological features. The hemipenis of *T. stejnegeri* is short, thick, not deeply forked, and bears heavy spines. The hemipenis of *T. albolabris* is longer, more slender, more deeply lobed, and is without spines. The cloaca of a female *T. stejnegeri* is shorter, not so deeply lobed, and has much thicker walls than that of *T. albolabris*. It seems likely that the long spines on the hemipenis of *T. stejnegeri* might damage the thin-walled cloaca of *T. albolabris* during copulation, while the slenderness and lack of spines of the hemipenis of *T. albolabris* might prevent its being held in place in the cloaca of *T. stejnegeri*. It is not known whether these two species do attempt to crossbreed in nature, but it may well be that here the differences in genital structure are the primary isolating mechanism.

Genetic Incompatibility

The accumulation of genetic differences in discontinuous populations may lead eventually to the development of genetic incompatibility between the two forms. There are many gradations of genetic incompatibility. Individuals of the two populations may be able to crossbreed to produce hybrid offspring that survive to maturity but are unable to reproduce successfully. The hybrids may show a lessened degree of viability so that few or none survive to maturity. The zygotes may fail to develop beyond the early stages, or there may be no embryonic development at all.

The widely distributed Leopard Frog (*Rana pipiens*) ranges from Canada to Central America and from the Atlantic to the Rocky Mountains. In different parts of the range, natural selection has brought about genetic reconstruction of the population to modify the physiological adaptations of the embryos. These embryos are narrowly adapted to the environmental conditions under which they normally develop. They differ in temperature tolerance, rate of development, and temperature coefficient of development. When individuals from adjoining parts of the range are crossed, the embryos develop in a perfectly normal manner, but the greater the geographic distance, the greater the percentage of abnormalities in the embryos. Vermont males crossed with New Jersey females produce perfectly normal embryos, whereas embryos of the reciprocal cross are either normal or very slightly abnormal. When Vermont females are crossed with males from Englewood, Florida, there is a marked retardation in the rate of development and the head of the embryo is greatly enlarged. The reciprocal cross produces embryos that show marked retardation in development and extreme reduction in the size of the head. Also, there is a high percentage of mortality and hence a high degree of hybrid inviability between these two populations. When even more distant populations of the species are crossed, the percentages of abnormal and inviable offspring increase and cellular disintegration is apt to begin quite early in embryonic development. Crosses between R. *pipiens* from Wisconsin and from Tamaulipas, Mexico, fail to produce any offspring that survive to the time of metamorphosis.

Crosses between species belonging to the Wood Frog group have produced results ranging from the production of hybrids that reach maturity to complete sterility. This group comprises a chain of species and subspecies extending from western Europe across northern Asia, through Alaska and Canada, to the eastern coast of North America. Crosses of two of the European species (*Rana temporaria* and *R. arvalis*) produce hybrids that survive to maturity, as do crosses of two Japanese forms (*R. japonica* and *R. temporaria ornativentris*). In crosses between *R. temporaria* males from western Europe and *R. sylvatica* females from eastern North America, development stops at the late blastula stage, and in the reciprocal cross there is no embryonic development at all. Apparently, genetic incompatibility between the two end species of the chain has reached the stage of causing complete sterility.

On the other hand, although hybrid viability frequently decreases in a single species with geographic distance, complete hybrid viability may exist between quite distinct, sympatric species. In eastern North America, *Rana pipiens* and *R. palustris* (the Pickerel Frog) produce hybrids that can be carried to transformation, although these species do not normally attempt to mate in nature. This is also true of the sympatric *Rana temporaria* and *R. arvalis* mentioned above. These species maintain their distinctness, not through genetic incompatibility, but through other isolating mechanisms.

When genetic incompatibility has developed between two forms and they come to occupy the same or contiguous areas, the individuals of one form that do not attempt to breed with individuals of the other will be the ones whose offspring will survive. Therefore, natural selection may be expected to bring about a parallel development of isolating mechanisms that inhibit attempts to cross.

In conclusion, two points about isolating mechanisms should be emphasized. The first is that probably no single mechanism ever works alone; in the separation of any pair of species several isolating mechanisms always seem to play a part. The second point is that isolating mechanisms develop through genetic reconstruction of populations initially separated ecologically or geographically, rather than that isolating mechanisms cause discontinuities to develop in a single continuous population.

Collateral Reading and General Reference

Blair, W. F. (editor). *Vertebrate Speciation*. Austin, Texas: University of Texas Press, 1961.

Dobzhansky, T. *Genetics and the Origin of Species*. New York: Columbia University Press, 1937. (Emphasizes the genetic basis of speciation. All of the books on this list are general works, but most of the principles discussed in them apply to the herptiles.)

Huxley, J. *Evolution: The Modern Synthesis*. New York: Harper Brothers, 1942. (A very comprehensive synthesis of the many, complex, and interrelated factors involved in speciation.)

Matthey, R. *Les Chromosomes des Vertebres*. Lausanne: F. Rouge, 1949. (Contains much information about the karyotypes of the herptiles.)

Mayr, E. *Systematics and the Origin of Species*. New York: Columbia University Press, 1942. (A companion volume to that of Dobzhansky, discussing both the contributions of systematic studies to an understanding of evolutionary processes, and the elucidation of systematic problems by modern theories of the mechanisms of speciation.)

White, M. J. D. *Animal Cytology and Evolution*. Cambridge: Cambridge University Press, 1945. (Deals with the structure and evolution of chromosomes and their bearing on evolutionary problems.)

GEOGRAPHIC
DISTRIBUTION

SINCE MAN FIRST BEGAN to explore the world around him, he has been aware that different kinds of animals live in different regions. Crocodiles are found in the River Nile but not in the Thames. Many snakes live in Europe but there are none in Ireland. So long as the doctrine of Special Creation held sway, it was easy to explain these facts by saying that each region had its own fauna, created especially for it and adapted to it. But this explanation was never really satisfactory. For one thing, there is too much overlap. Not all animals are restricted to one region; some are spread over several. Many of the animals the early explorers found in the New World were different from anything they had ever seen, but others were strikingly similar to those they had known at home. Then it became obvious that animals may be well adapted to regions where they do not naturally occur. The horses the Spanish Conquistadors brought to America throve in the new environment. Florida is obviously a fine place for the Cuban Greenhouse Frog (*Eleutherodactylus ricordi planirostris*). It was first reported from Key West in 1863 and 80 years later had spread almost to the northern border of the state. When the true nature of fossils was finally recognized, and their orderly succession became apparent, men realized that some animals once lived in regions where they are no longer found and that many forms have become extinct. Dinosaurs no longer exist, but they once roamed every continent. It became necessary to postulate a whole series of special creations, with the next to the last one presumably wiped out by the Biblical Flood. The doctrine eventually became absurd.

It was Darwin's observations on the distribution of animals in South

America and the nearby Galápagos Islands that first turned his thoughts to the idea of evolution. Now the two studies supplement and reinforce one another. The pattern of distribution of animals is one of the strongest arguments for evolution and offers many clues to the course evolution has taken. Conversely, to understand the present distribution of animals we must take into account not only their ecological requirements but also their evolutionary history. It is to be hoped that the correlation of the geographic distribution and evolutionary history of some groups that are well represented in the fossil record will enable us to form sound ideas concerning the evolution of other groups by an inspection of their geographic distribution.

FACTORS IN GEOGRAPHIC DISTRIBUTION

The geographic distribution of any group of animals is the result of the interplay of two sets of factors, extrinsic and intrinsic. They may be tabulated as follows:

I. Extrinsic factors

    A. Distribution of favorable environments
    B. Changes in environments through geologic time

        1. Climatic
        2. Biotic

    C. Formation of highways permitting dispersal or of barriers to dispersal

II. Intrinsic factors

    A. Physiological requirements of group
    B. Time and place of origin
    C. Potential rate of spread

        1. Biotic potential
        2. Vagility

    D. Genetic plasticity of group

Extrinsic Factors

It is obvious that, for a group of animals to occur in a region, environments must be available that are suited to the needs of that particular group. We do not expect to find the moisture-loving salamanders in a desert. *Salamandra salamandra* has certainly been able to reach northern Africa, but it has not been able to extend its range out into the Sahara. The cold arctic regions are not suited to the ectothermic herptiles and few of them are found there. The True Frogs (*Rana*) reach the Arctic in both the Old and New Worlds and a salamander (*Hynobius keyserlingi*), a lizard (*Lacerta vivipara*), and a snake (*Vipera berus*) extend that far north in the Old World, and that is all. A favorable environment implies not only suitable physical features but also favorable biotic conditions. There must be a nice balance between predator and prey species, hosts and parasites, competitors and food.

However, most environments are stable only over relatively short periods of time in the geologic sense. One of the best known of the changes that have taken place is the warming up of the northern hemisphere in the last ten thousand years. This is evidenced by the retreat of the glaciers since the close of the Ice Age. The process is apparently continuing to the present day and many groups of animals are still actively extending their ranges northward. Further back in geologic time, the coal measures in Pennsylvania show that semitropical swamps once flourished in what is now a hilly, well-drained, temperate region.

Climatic factors of the environment are not the only ones that change, biotic factors are also constantly shifting since the forces of evolution are continually at work on all forms of life. New sources of food become available to animals able to take advantage of them, new enemies appear which must be evaded. Perhaps most important of all, new and better adapted competitors for food and breeding sites either move into the area or evolve within it.

Although the major continental land masses have probably remained relatively constant, at least since the appearance of the first amphibians in the Paleozoic, there have been many changes in the connections between them. South America was cut off from North America by an arm of the sea for the greater part of the Cenozoic Era. During this time distinct faunas evolved in the two regions. When the land connection was reestablished during the Pliocene, it became possible for North American forms to invade South America and vice versa. The mingling has been very incomplete and the two faunas are still essentially different. Thus only one group of salamanders (*Bolitoglossa* and its allies of the family Plethodontidae) has invaded South America. In contrast, the North American continent has apparently been connected in the past with Asia across the Bering Straits at times when the climate was sufficiently mild so that this land bridge proved a suitable highway for the dispersal of many groups of amphibians and reptiles. The fauna of temperate North America resembles that of Asia more closely than it does that of South America.

Obviously, a given topographic feature may serve as a highway of dispersal for some forms and as a barrier to others, depending on the physiological requirements of the groups involved. Broad lowland valleys are barriers to salamanders adapted to dwelling on mountain tops but may serve as highways to other forms such as toads.

Intrinsic Factors

Even closely related forms often show great differences in their physiological requirements. Furthermore, some animals have broad ecological tolerance and are able to adapt themselves to conditions over a wide area while others are very limited in ecological tolerance and hence are restricted to a narrow range. This sort of difference is probably reflected in the distribution of two of the North American Rat Snakes. The ecologically adaptable *Elaphe o. obsoleta* ranges from Ontario and northern New England south to Georgia and west to Minnesota and Texas while *E. subocularis* is limited in range to the arid region of Trans-Pecos Texas, southern New Mexico, and adjacent Coahuila.

Besides the physiological requirements of the animal, which determine what environments it may occupy and what highways of dispersal are available to it, the time and place of origin of a group play an important part in determining its geographic distribution. If a group arises in a region shut off from an adjacent region by some barrier, it will not be able to spread into that region even though there may be environments there well suited to its needs. The family Plethodontidae, which undoubtedly originated in eastern North America in the early Cenozoic, is probably not widespread in South America simply because only recently, in the geologic sense, has a passageway been opened.

If a group is of very recent origin, it may not have had time to spread very far from its center of origin. How fast a group will spread depends in part on its biotic potential. Animals capable of producing large numbers of offspring in a relatively short time will, other things being equal, be able to occupy new areas more rapidly than forms with a low rate of reproduction. The vagility of the species, that is, the inherent power of movement of the individuals, may also have its effect on rate of spread. Both of these factors are probably relatively minor since many forms with low biotic potentials and limited vagility have been able to occupy large areas of the earth's surface.

Finally the genetic plasticity of the group will determine whether it will be able to occupy new environments, whether it can adapt itself to changes in the environment *in situ*, or whether it must either follow receding belts of its old environment or become extinct.

PATTERNS OF DISTRIBUTION

During their evolutionary history, animal groups pass through various stages. When a group first arises, it is small in number and occupies a limited area. It then goes through an expanding phase in which it spreads out to occupy such territory as is available to it as determined by the factors outlined above. This is usually followed by a contracting phase in which the population is reduced and eventually extirpated over much of its former range. Since different groups have arisen at different times during the past and since, furthermore, they differ in the rates at which they pass through these stages, we must expect them to show differences in distributional pattern.

Expanding Populations

It is axiomatic that a population of animals must be more or less adapted to the environment in which it is found or perish. But as time goes on this population, or at least a part of it, is exposed to the harsh influences of new environments to which it is not adapted. This may come about in two ways.

Most animals are capable of producing many more offspring than could be expected to survive within the limits of the area occupied by the parent population. This means that overcrowding is an ever-recurring phenomenon in any but a decadent population. The animals near the periphery of the range are constantly being pushed into new and unfavorable environments. Undoubtedly, most of them perish. Stocks that have enough genetic plasticity are sometimes able to make the change so that their descendants become adapted to the new conditions and hence develop into incipient discrete populations. Under this process we have the primitive types occurring toward the center of the geographic range while specialized types develop at the periphery. This is the so-called rim-fire pattern of distribution.

The frog genus *Eleutherodactylus* is widespread in northern South America and the species in this area seem to be relatively stable. At the periphery of the range, in the West Indies, the group is undergoing rapid speciation. Thus, on the relatively small island of Jamaica, besides two recently introduced forms, there are fourteen endemic species of *Eleutherodactylus*. (Species endemic to a region are ones found only in that particular region.)

These fourteen apparently all evolved on Jamaica from three, or at most four, stocks that reached the island from the mainland.

Since the world's environments are constantly, albeit at times very slowly, changing, this pattern of distribution must most commonly occur in relatively small areas and over relatively short periods of time.

The second way in which a population of animals may be exposed to a new environment is through changes in the climatic and biotic conditions within the area in which it lives. This results in a distribution pattern quite different from the one discussed above. Animals near the periphery of the range may be able to follow receding belts where the old conditions are still extant. They may thus remain unchanged, while the forms toward the center of origin must become adapted to the new conditions or perish. Since most environmental changes over large areas take place but slowly this is often possible. Here we have specialization taking place near the center of origin of the group, with the primitive types located at the periphery—the center-fire pattern of distribution. This process applies principally to larger, more inclusive groups occupying extensive areas over long periods of time.

The widespread frog family Ranidae shows this type of distribution. Africa seems to have been a center of differentiation for the group. Six of the eight subfamilies are found there and four are found nowhere else. The rather unspecialized genus *Rana* reaches Europe, North and South America, and Australia.

Expanding populations generally exhibit continuity of range. When this is so, the lines of continuous distribution should lead back toward the center of origin of the group as a whole. Thus, if the range of an animal group (species, genus, family) is continuous and we find the specialized types at the periphery (in young, small groups) or concentrated near the center (in older, larger groups) we have good indications that the center of origin of the group is somewhere near the geographic center (that is, the center of available migration routes) of the group.

Relict Populations

Since the history of the fauna of the earth has been one of extinction as well as of evolution, we must expect to find in addition to expanding populations many examples of decadent and relict stocks. Although the causes and rates of extinction are not completely understood, apparently subservience to later, better-adapted types plays at least as important a role as rigorous changes in the physical environment. Thus a population that originated and spread from some center of origin may well be eliminated by a later, more successful population originating at the same center. Hence, centers of origin may also become centers of extinction.

Except for certain island faunas, perhaps the best clue to a decadent population is discontinuity of range. Where there are gaps in the distribution, it means either that a stock has been introduced into a distant area or that it spread there under its own power and the links connecting it to the parent stock have been wiped out. The alligators, with one form in the southeastern United States and one in southeastern Asia, have a typical relict distribution. Apparently the alligators living today are but remnants of an earlier, widespread stock. The fossil record confirms this.

Thus, where we find a group divided into isolated stocks, there is a strong probability that we are dealing with a receding rather than an expanding population and that these isolated stocks are relicts. Since the center of extinction of a group may be anywhere within its expanded range, it follows that in these receding populations the geographic center of the present distribution need not be near the center of origin of the group. In fact, a receding population may well be extinct at the place of origin. Here we must lean heavily upon the fossil record to determine the center of origin.

Waif Populations

A waif population is one whose ancestors reached an area (usually an island) as stray or castaway individuals rather than through the normal dispersal of an expanding population. In times of flood, uprooted trees or segments of the river bank may be swept downstream and carried far out to sea. They may be caught by ocean currents and stranded eventually on the shores of a distant island. Such a natural raft may harbor a clutch of eggs, a pregnant female, or a pair of individuals of the same species. If they survive the hazards of the journey, they may be able to establish a population on the island.

It is not always easy to tell whether a population is waif or whether it reached the island at a time when the island was still connected to the mainland. In either case, speciation may have occurred so that the population differs from any now found on the mainland. Islands that have been formed entirely by volcanic action or that have been completely submerged since their last previous connection with another land mass must *ipso facto* be populated by waifs.

In the past few thousand years, man has, either accidentally or purposely, carried many animals from one part of the globe to another. It is customary, though, to speak of populations arising from animals transported by human agencies as introduced rather than waif species.

The island of Bermuda is inhabited by a species of lizard, *Eumeces longirostris*, which differs greatly from all other species of the genus and is found in no other place. It is undoubtedly waif. The other herptiles of the island, another lizard and three frogs, have all been introduced by man.

ZOOGEOGRAPHIC REALMS

With the development of a dynamic approach to zoogeography, several attempts have been made to divide the earth's surface into zoogeographic realms, broad areas characterized by general resemblances of the fauna within each realm and general differences between the faunas of one realm and another. One of the first of these atttempts was by an ornithologist, P. O. Sclater, who, in 1858, recognized six major realms. His divisions were so carefully selected that they still form the basis of our divisions today. His terminology, however, has been somewhat modified. A recent and acceptable nomenclature for these regions follows (see also Fig. 11-1).

Arctogaea:	Holarctic Region:	Palearctic subregion
		Nearctic subregion
	Ethiopian Region:	African subregion
		Madagascar subregion
	Oriental or Indian Region	
Notogaea:	Australian Region	
	New Zealand Region	
	Oceanic Islands Region	
Neogaea:	Neotropical Region	

The Holarctic Region includes all of Europe and its islands, the northern belt of Africa, Asia north of the Himalayas, and all of North America north of the West Indies and an ill-defined zone in Mexico. It is readily and logically separable into the Palearctic subregion, the Old World portion of the Holarctic, and the Nearctic subregion, the New World portion of the Holarctic.

Madagascar, the Arabian peninsula, and all of Africa except the belt north of the Sahara form the Ethiopian Region. It is divided into a Madagascar subregion for that island alone and an African subregion.

The Oriental Region leads into the Australian Region by way of an archipelago. The islands of this archipelago form a series of stepping stones, so to speak, between the two. It is a matter of dispute between which two islands the boundary line should be drawn, if indeed the boundary is definable as a single line. The Oriental Region, as here defined, includes south-

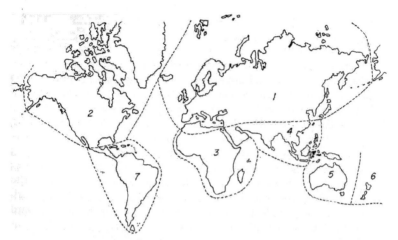

FIG. 11-1. Zoogeographic regions and subregions. 1. Palearctic Subregion; 2. Nearctic Subregion; 3. Ethiopian Region; 4. Oriental or Indian Region; 5. Australian Region; 6. New Zealand Region; 7. Neotropical Region. [Modified after Beaufort.]

ern Asia south of the Himalayas and extends to the southeast through Ceylon, the Philippines, Sumatra, Borneo, and Java.

The Australian Region then comprises the islands south and east of those listed immediately above to and including Australia and the islands on its continental shelf.

North and South Islands of New Zealand and the smaller islands in the vicinity comprise the New Zealand Region.

The scattered islands of the Pacific are recognized as the Oceanic Islands Region.

Finally, the West Indies and Central and South America make up the Neotropical Region.

Although it is best not to attempt to define these regions too sharply, they are basically useful and the names of many of them are so firmly entrenched in the literature that we must have an understanding of them. Much difficulty has been caused by attempts to apply the names of the regions to their faunas. Since none of the regions has a sharply limited, uniform fauna, such attempts are at the outset bound to cause confusion. For example, it has been shown that the herpetological fauna of the Nearctic Region consists of a modern, circumpolar element, an "Old Northern" element characteristic of the southern United States, and a South American element representing a recent invasion from the south. Obviously we can speak of the region as Nearctic but not the fauna.

ORIGIN OF THE FAUNA

There have been several attempts to select one area or another as the major center of origin and dispersal for present-day forms. However, there seems to be no inherent reason why all or most of the modern families should have originated in a single region. Just as they arose at different times in the past, so also did they probably arise in different places. The majority of them may have appeared first on the great Holarctic land mass, simply because it is the largest of the major land areas. The primitive and relict stocks in the southern hemisphere are thought to have been derived from populations formerly occupying the nearly continuous land masses to the north, from which they have been eliminated by the later, more advanced types and modern forms; thus remnant populations persist toward the ends of the continental land masses that extend southward like great peninsulas (Fig. 11-2). Many amphibian and reptile stocks exhibit this pattern of distribution. The side-necked turtles at present live only in South America, Africa and Madagascar, and Australia and New Guinea, although they are known from the late Mesozoic of Europe, North America, and South America. Among the amphibians the Pipidae are limited to Africa south of the Sahara and to South America. Nonetheless, it is not likely that all of our modern herptiles are holarctic in origin. There have unquestionably been many secondary centers of evolution and some of them have undoubtedly been large enough and in existence long enough to have permitted the evolution of new genera and families of animals. Many genera of Hylidae have evolved in northern South America, and many genera of Ranidae in Africa.

GEOGRAPHIC DISTRIBUTION OF
LIVING AMPHIBIANS AND REPTILES

Despite their relative scarcity as fossils, in some respects amphibians and reptiles are among the better animals to use in studies of geographic distribution. Like the vast majority of the world's animals they are ectothermic. The principles of distribution that they exemplify are thus apt to be of wider applicability than those shown by the better known endothermic birds and mammals. They are limited in vagility as compared to the winged insects and birds or the widely ranging larger mammals. The outlines of their primary patterns of distribution are thus less apt to be blurred by rapid, wide, and essentially random dispersal. Because of their unshelled eggs and unprotected skins, amphibians are ecologically bound to regions where fresh water is available. They are quite intolerant of sea water. We

NORTH
°
POLE

FIG. 11-2.
Diagram of the major land masses of the world, showing how the continental masses extend southward from the north polar region like great peninsulas.

195

can be reasonably sure then that they have not fortuitously crossed extensive areas of the sea or arid desert lands *en masse*. Where we have essentially similar faunas on either side of such a barrier we can safely assume that they reached those regions before that barrier formed.

Apoda

Although no fossil caecilians are known, the widely discontinuous distribution of the present day forms indicates that they are relicts of a once widespread stock. The group is essentially a tropicopolitan one which occupies four land masses: tropical America south to Buenos Aires; tropical Africa; the Seychelles Islands in the Indian Ocean; and the Oriental Region from Ceylon and India to Java, Borneo, and the southern Philippines. Caecilians are absent from the Lesser Sundas, Celebes, and the Australian region of the Orient, from Madagascar, and from all of the West Indies except Trinidad. Of the approximately seventeen genera known, none is found in more than one of the major land areas occupied by the group.

Trachystomata

Modern trachystomes are found only in southeastern and central United States and northeastern Mexico, but fossil representatives of the order are known from the Cretaceous of Wyoming. The group was thus once much more widely spread in North America and is obviously relict today.

Caudata

The most striking characteristic of salamander distribution is the discontinuity of the major groups, reflecting wholesale extinction and the relict nature of the group as a whole. Of the seven recent families, all but two show more or less discontinuous distribution. Thus in the family Cryptobranchidae, one genus, *Megalobatrachus*, is found in China and Japan and the other, *Cryptobranchus*, in the eastern United States. While the caecilians are tropical, the salamanders are a north temperate group and have invaded the tropics only in Central America and northern South America; this invasion has been made only by one group (*Bolitoglossa* and its allies of the Plethodontidae). At present the order is centered in North America. Six of the seven families are found there, with only the Hynobiidae being absent. Two families, Ambystomatidae and Amphiumidae, are restricted to North America.

Anura

The flourishing group of modern amphibians is that of the frogs and toads. They are abundant on all continents; the Oriental Region and South America are very rich in them. The three most primitive families all show the typical disjunct distribution of relict forms; the Leiopelmidae are found in northwestern North America and New Zealand, the Pipidae in South America and Africa, and the Discoglossidae in temperate Europe and Asia and the Philippines. It is interesting that these primitive frogs do not occur in Australia. The frogs now found in that region, members of the families Leptodactylidae, Hylidae, Ranidae, Rhacophoridae, and Microhylidae, are advanced types known to be adept at "island hopping" (that is, they are easily transported from one island to another). The presence of a leiopelmid on New Zealand (the only frog known from there) suggests that this family once occurred in Australia as well; perhaps it could not withstand the competition from the later arriving island hoppers. The most successful family of frogs is the Ranidae, members of which are found on all major land masses except Greenland and New Zealand.

Testudinata

The turtles are a rather ancient and conservative group; many of the present families show discontinuities in distribution. Our knowledge of how they got to their present places of abode is better than for many of the other groups of herptiles, simply because they leave a better fossil record. Thus the Carettochelyidae are at present found only in New Guinea but their fossils are known from Europe, Asia, and North America. Evidently the recent form (*Carettochelys insculpta*) represents a much reduced relict of a once wide-spread stock. The primitive side-necked turtles show the typical disjunct distribution of relict forms: the Pelomedusidae are found in Africa, Madagascar, and South America; the Chelidae in Australia, New Guinea, and South America. The most flourishing and wide-spread family is the Testudinidae which includes the pond turtles and land tortoises. It is almost cosmopolitan in distribution, only Australia being without representatives.

Rhynchocephalia

There is only one living form of this ancient order. *Sphenodon* is now known only from a few islands off the coast of New Zealand but the family goes back to the Lower Triassic and is known in fossil form from Africa, Europe, and North America.

Lacertilia

Although lizards are particularly abundant in the tropics, they occur in suitable places on all the continents and one, *Lacerta vivipara*, reaches well above the Arctic Circle in Scandinavia. One striking thing about the distribution of the families of lizards is that there apparently has been less exchange between the faunas of the Old World and the New World than has taken place in the other major groups of herptiles. Of 23 families, only 8 are found in both, 9 are restricted to the Old World and 6 to the New. Interestingly enough, the most primitive infraorder, the Gekkota, is also one of the most widely distributed. Members of it are found in tropical regions throughout the world. Apparently these lizards are peculiarly liable to fortuitous introduction; the original distribution of some of them may never be worked out. Some lizard families that have restricted distributions at present are apparently relicts. Thus the monitor lizards of the family Varanidae, which are now found in the tropics and some warm temperate regions of Africa, Asia, and Australia, are known as fossils from North America and Europe. On the other hand, some of the families of lizards with limited distribution are probably quite recent and have not yet been able to spread far from their centers of origin.

Serpentes

The geographic distribution of the snakes is, in a sense, the most difficult of all to analyze. This is because the family Colubridae, which contains about two-thirds of the world's living snakes, is really not well enough known systematically. Although the genera and species have been defined fairly adequately, the relationships of the genera within the family have not been satisfactorily worked out.

The snakes, like the lizards, are widely distributed and occur in habitable places on all continents. They are absent from New Zealand and Ireland. They are most abundant in the tropical regions of the Old World. All of the 10 families currently recognized, and most of the subfamilies, are found in the Oriental Region and 3 are restricted to it. Seven families occur in the New World but none is found there exclusively.

Crocodilia

The Crocodilia are mainly tropicopolitan in distribution, but one genus, *Alligator*, inhabits the north temperate zone, occurring in the southeastern United States and southeastern Asia.

Collateral Reading and General Reference

Beaufort, L. F. de. *Zoogeography of the land and inland waters.* New York: Macmillan, 1951. (A sound, descriptive, regional zoogeography.)

Darlington, P. J. Jr. *Zoogeography: The Geographical Distribution of Animals.* New York: John Wiley, 1957. (The modern definitive work on zoogeography.)

Dunn, E. R. "The herpetological fauna of the Americas." *Copeia,* 1931, no. 3, pp. 106–119. (A classic paper on reptile distribution in the New World.)

Matthew, W. D. "Climate and Evolution." *Annals of the New York Academy of Sciences,* Vol. 24, pp. 171–318; reprinted as *Special Publication, New York Academy of Sciences,* Vol. 1, 1939. (Modern zoogeography begins with Matthew's great work. It should be familiar to all who are interested in the distribution of animals.)

Simpson, G. G. "Turtles and the origin of the fauna of Latin America." *American Journal of Science,* no. 241, pp. 413–429, 1943. (A stimulating zoogeographic paper by a paleozoologist.)

Smith, M. A. *The Fauna of British India.* Vols. I–III. London: Taylor and Francis, 1931–1943.

Wallace, A. R. *The Geographical Distribution of Animals.* London: Macmillan, 1876. (The original definitive work on zoogeography.)

CAECILIANS,
TRACHYSTOMES,
AND SALAMANDERS

CAECILIANS, SIRENS, AND SALAMANDERS are treated together in this chapter simply as a matter of convenience. They are mostly small, inconspicuous, secretive animals, limited in number, both of kinds and of individuals, and restricted in distribution. So it is not surprising that they are in many ways the least familiar of the herptiles. The tropical, burrowing caecilians are completely unknown to most nonbiologists and are little more than names and pictures in a book to many biologists. There are only two modern genera of trachystomes, both very limited in distribution. The salamanders are somewhat more numerous, and because of their availability in the north temperate zone they have been the subjects of many excellent studies on life history, evolution, and ecology. *Necturus*, of course, is familiar to most comparative anatomy students. Much of our knowledge of how organs are differentiated during development has come from experimental transplanting of tissues on the embryos and early larvae of *Ambystoma* and *Triturus*. Otherwise, the salamanders have been almost ignored by experimental biologists and laymen continually confuse them with lizards.

ORDER APODA (GYMNOPHIONA)

A caecilian is a slim, wormlike creature with no limbs or limb girdles and practically no tail. Most are not more than 240 to 300 mm. in length. The

200

vent is close to the posterior end of the body on the ventral side; the eyes are minute, buried in the skin, and without eyelids; the skull is compact; and the intestine is not differentiated into large and small portions. Adults lack gills or gill slits. Fertilization is internal and the cloaca of the male is modified to form a protrusible copulatory organ. The more primitive genera have tiny, dermal scales imbedded in the skin, apparently a heritage from the early, scaled amphibians of the Carboniferous. All caecilians are included in a single family.

Family Caecilidae

Caecilians are primarily forest animals. They have been reported from savanna areas, but only along rivers bordered by patches of forests. One form, *Typhlonectes*, is an aquatic river dweller. All the rest are probably terrestrial and fossorial, living in burrows in damp earth. Snakes, which can hunt caecilians in their burrows, seem to be their main predators.

Some caecilians, such as *Rhinatrema*, *Siphonops*, *Hypogeophis*, and *Caecilia*, are oviparous, but others, including *Gymnopis*, *Chthonerpeton*, *Typhlonectes*, *Geotrypetes*, and *Schistometopum*, are viviparous, giving birth to living young. *Ichthyophis*, a native of Asia, breeds in the spring. The female prepares a burrow in moist ground close to running water. She coils her body about the 20 or more relatively large-yolked eggs, and guards them zealously from predaceous snakes and lizards. The eggs absorb water and gradually swell until they are about double their original size. The larva at hatching weighs approximately four times more than the new-laid egg. External gills are present at first but are lost soon after the larva hatches. The young go through a long aquatic stage before they metamorphose into burrowing, terrestrial adults that would drown if kept under water.

The genus *Rhinatrema* of northern South America also lays eggs from which hatch aquatic larvae with external gills. But *Gymnopis* and *Geotrypetes* retain the eggs in the oviduct where hatching takes place. The wall of the oviduct is provided with compound oil glands and the larvae subsist by eating this wall with its included oil droplets. They metamorphose before they are born so that the newborn young are replicas of the adults.

ORDER TRACHYSTOMATA

This is another small order sharply distinct from all other amphibians. Its members are aquatic, permanent larvae and develop few adult characteristics. They lack maxillaries and cloacal glands. The eyes are tiny, the front legs are minute, and the hind legs are lacking entirely. Both jaws are sheathed with horn. A Jacobson's organ, lungs, and three pairs of external gills are present. There is only one family in the order.

Family Sirenidae

The family Sirenidae includes two recent genera, *Siren* with two species and *Pseudobranchus* with one. It is now restricted to southeastern and central United States and extreme northeastern Mexico. *Pseudobranchus* is a slender little animal having only three toes on each foot and a single pair of gill slits. Members of the genus *Siren* have four toes on each foot and three gill slits. They are much longer and more heavily built than *Pseudo-*

FIG. 12-1. The largest living trachystome, *Siren lacertina*, an abundant amphibian in the southeastern United States.

branchus. Indeed, *S. lacertina,* which reaches a length of 90 cm., is one of the largest of the amphibians.

Nothing is known of the mating habits of the trachystomes. They have repeatedly been said to practice external fertilization. But this is simply because no one has demonstrated either the production of spermatophores or the presence of spermathecae (pockets of sperm storage) in the cloaca of the female. Nonetheless, we remain unconvinced that the Sirenidae do not have internal fertilization. A mature female of the genus *Pseudobranchus* may have over 100 fully developed and pigmented eggs apparently ready for deposition at one time. The eggs, when laid, are attached singly to the roots of water plants and may be spaced as much as 1, 2, or 3 m. apart. Often they are so widely scattered that less than a dozen eggs can be collected in an entire afternoon. It is difficult to see how such eggs could be fertilized after they are laid.

FIG. 12-2.
The egg and newly hatched young of the Narrow-striped Dwarf Siren, *Pseudobranchus striatus axanthus.* The egg is attached to a water hyacinth rootlet. [After Goin.]

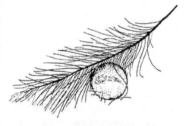

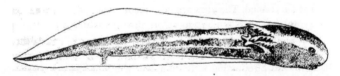

The eggs of *Pseudobranchus* vary from about 7 to 9 mm. in diameter. Although the exact developmental time is not known, it must be several weeks since a series of eggs collected in the neural groove stage did not hatch until 17 days later. The newly hatched larvae range from about 14 to 16 mm. in total length. There are no balancers, the toes are differentiated, and a well-developed dorsal fin extends from the base of the head to the tip of the tail. There is perhaps less evidence of metamorphosis in the Sirenidae than in any of the perennibranch salamanders, but *Siren* does develop an adult-type skin.

ORDER CAUDATA (URODELA)

As the name of the order indicates, a salamander retains its tail throughout life instead of losing it at metamorphosis as a frog does. The head and trunk regions are distinct, not fused as they are in a frog. Two pairs of weak legs and a primitive, poorly developed sternum are present. Fertilization is either external or internal by means of spermatophores. Most salamanders are oviparous. The larvae, which closely resemble the adults, have true teeth in both jaws. Except for the perennibranchs, permanent larval types that retain the gills as adults, the lateral line system is lost at metamorphosis.

The salamanders of today do not show the extensive adaptive radiation displayed by many of the other tetrapods. Some are terrestrial, some aquatic, and some may live in either environment. A few have become somewhat arboreal and a number are fairly efficient burrowers, but this is about the limit of their capabilities. For the most part, salamanders are simply creatures that must stay in or on moist ground, or entirely in water.

The living salamanders are divided into seven families, comprising about 280 species. These families fall into four natural groups which are here called suborders. The first three, Cryptobranchoidea, Ambystomatoidea, and Salamandroidea, seem to have descended from a common ancestor, but the relationship of the fourth group, the Proteida, to the others is obscure.

Suborder Cryptobranchoidea

These are the most primitive of the salamanders and the only ones known to have external fertilization. Other salamanders have a complex of glands in the cloaca which contribute to the formation of the spermatophores. The cryptobranchoids possess only one type of cloacal gland and no spermatophores are formed. The eggs are laid in gelatinous sacs.

Cryptobranchoids are also more primitive structurally than the other salamanders. The earliest amphibians resembled the fishes in having many different bones in the skull. In the evolution of the group, the tendency has been toward a reduction in the number of bones through loss or fusion. In the cryptobranchoids, two of the bones of the lower jaw, the angular and prearticular, are still separate. They are fused in the higher salamanders.

There are only two families in the suborder, one confined to Asia, the other found both in Asia and the eastern United States.

Family Hynobiidae. The Hynobiidae, Asiatic land salamanders, undergo a more complete metamorphosis than the members of the second family in the suborder. They develop eyelids, nonlarval teeth, and other metamorphic characters. The family is concentrated in eastern Asia and the adjacent islands. There is one widespread, primitive genus, *Hynobius,* whose range extends from Japan to western Asia. (*Hynobius keyserlingi* is sometimes placed in a separate genus, *Salamandrella.*) The other four genera were apparently all independently derived from *Hynobius* and are all included within its geographic range. *Pachypalaminus* and *Batrachuperus* have well-developed, horny epidermal pads on the soles, palms, and digits. *Batrachuperus* differs from both *Hynobius* and *Pachypalaminus* in having the vomerine teeth separated into two small, isolated patches rather than in a V-shaped series. *Ranodon* and *Onychodactylus* are mountain stream forms. Salamanders living in mountain torrents must keep themselves from being carried downstream by the current. Air-filled lungs make an animal buoyant and increase the chances of its being swept away. Moreover, the water in such streams is cool and contains much oxygen. A salamander living in cool water has a low body temperature, and consequently a low metabolic rate. It does not need as much oxygen as a warm water form, and is able to get what it needs from the oxygen-rich water through cutaneous respiration. In such an environment, the disadvantages of lungs outweigh their advantages, and they tend to be reduced or absent. They are small in *Ranodon*. *Onychodactylus* has become completely lungless, thus paralleling the more familiar plethodontid salamanders of the United States.

Hynobiids have not developed the rather elaborate courtship behavior shown by other salamanders. The male is stimulated to sexual activity by the extruding egg sacs, and his sole response to the female is apparently an attempt to push her away as he fertilizes the eggs. *Batrachuperus karlschmidti* is a common salamander of small mountain streams at elevations of about 1.8 to 4 km. in western China. The female attaches the egg case under or to the side of a large stone in flowing water. The end attached to the stone is flat and sticky and the body is a cylindrical tube, largest in the middle and smaller toward the transparent free end. The free end is covered with a smooth, rather delicate cuplike cap. This cap is forced off by the movement of the fully developed embryos, which free themselves through the hole thus formed. The individual egg cases contain from 7 to 12 eggs or developing embryos. Since as many as 45 eggs in the same stage of development have been taken from a single specimen, presumably each female must deposit 5 or 6 separate egg cases. The larvae are fairly typical salamander stream larvae.

Family Cryptobranchidae. This family contains the giant salamanders and hellbenders—squat, ungainly water animals that never completely metamorphose. In contrast to the hynobiids, the adults lack eyelids (as do all larval salamanders) and retain the larval teeth. Although they never leave the water, they do undergo a partial metamorphosis and adults of both New and Old World forms lose the gills. In the American Genus, *Cryptobranchus*, one gill slit remains open, whereas in *Megalobatrachus* of eastern Asia all

FIG. 12-3. The Hellbender of the eastern United States, *Cryptobranchus alleganiensis.*

are closed in the adult. The cryptobranchids also differ from the hynobiids in having flattened skulls from which certain bones (lacrimals and septo-maxillaries) have disappeared. *Megalobatrachus,* which reaches a length of 150 cm., is the largest of the salamanders, and indeed of the living amphibians. *Cryptobranchus* is smaller, generally being about 68 cm. long.

Although apparently once more extensive, this family is now represented only by these two genera. The hellbender of the eastern United States, *Cryptobranchus alleganiensis,* ranges from New York south to Georgia and west to the Ozarks. *Megalobatrachus* comprises two species, *M. japonicus* of Japan and *M. davidianus* of China.

Like the hynobiids, salamanders of the family Cryptobranchidae have external fertilization. *Cryptobranchus* mates in the late summer. At this season the male excavates a nest in the stream bottom beneath some large sheltering object, usually a flat rock. He allows females that have not deposited their eggs to enter the nest but drives spent females or other males away. The eggs are laid in long, rosary-like strings, one from each oviduct; these form a tangled mass at the bottom of the nest. As many as 450 eggs may be deposited by a single female and at times several females may lay in a single nest. As the eggs are deposited the male discharges a whitish, cloudy mass into the water. This consists of the seminal fluid and the secretions of the cloacal glands. While the eggs develop, the male often lies among them with his head guarding the opening of the nest. The incubation period lasts between 10 and 12 weeks and the young larvae hatch out at about 30 mm. in length. They lose their gills when about 125 mm. long at an age of approximately 18 months.

Suborder Ambystomatoidea

Although these salamanders are apparently descendants from some early hynobiid stock, they have diverged enough to be placed in a separate suborder. They are usually rather sturdily built, broad-headed salamanders, small to medium in size. Adults are generally terrestrial. There are three sets of cloacal glands in the male, spermatophores are formed, and fertilization is internal. The ambystomatids are more advanced than the hynobiids in having the angular fused with the prearticular in the lower jaw. Skeletal characters are also used to separate the ambystomatids from the Salamandridae. In the palatal region of the skull of the former, the prevomers are short and lack posterior processes extending back over the parasphenoids. Most ambystomatids have a crossbar between the posterior horns of the hyoid apparatus. The suborder contains only one family and is found only in North America.

Family Ambystomatidae. The most abundant and widely distributed genus, *Ambystoma*, ranges from eastern to western North America and comprises about 15 species. One of these, *A. tigrinum*, with its various races, is found from coast to coast. Two other genera of ambystomatids, *Dicamp-*

FIG. 12-4. The Barred Tiger Salamander, *Ambystoma tigrinum mavortium*, a typical ambystomatid salamander.

todon and *Rhyacotriton*, each with a single species, are limited in distribution to western North America. The primitive and partially neotenic *Rhyacosiredon*, comprising four species, is found only in the high mountains at the southern edge of the Mexican plateau. *Dicamptodon* is the largest of the land salamanders, reaching a length of 271 mm. The costal grooves along the sides, which in many salamanders mark the divisions between the segments of the trunk musculature (myotomes), are less well defined in *Dicamptodon* than in *Ambystoma*. The tail is more strongly compressed and thin-edged above. *Rhyacotriton* is a small, mountain stream form with vestigial lungs. Many salamanders have an ypsiloid (Y-shaped) cartilage extending forward from the pelvic girdle on the ventral side which helps control the shape of the inflated lungs. It is reduced in *Rhyacotriton*, as in most lungless salamanders.

The breeding habits of *Ambystoma jeffersonianum* are probably typical of the family. The adults migrate to the breeding ponds in early spring. Females usually outnumber males and must often bid for attention. There is a characteristic liebespiel before the spermatophore is deposited. The female lays the eggs in small, cylindrical masses each containing, on an average, about 16 eggs. These are attached to slender twigs or other objects below the surface in quiet pools. Since the egg complement of the female may total over 200, it takes a number of masses to complete egg deposition. The upper part of the egg is dark brown to black, a pigmentation characteristic of amphibian eggs laid in the open. Under field conditions, the incubation period ranges from about 30 to 45 days. The dark-colored hatchling is about 12 mm. in length, and has well-developed balancers; the forelegs are represented by elongate buds directed backwards, but the hind legs are not yet developed. The tail fin is continuous with the back fin which extends almost to the base of the head. Metamorphosis usually takes place two to four months after hatching, and the transforming young may be found from July through September.

Most other ambystomatids also lay their eggs in water in the early spring, but *Ambystoma opacum* lays its eggs on land in the fall. The female coils around them and protects them. The young, which have all the typical larval characteristics, hatch out when winter rains begin and then make their way into the water.

Ambystoma is notorious for its tendency toward neoteny. The well-known axolotls of Mexico were long considered a distinct genus of permanent larvae (*Siredon*) until it was found that, by altering their environment, they could be induced to metamorphose into perfectly typical, terrestrial ambystomas.

Suborder Salamandroidea

This is the dominant group of salamanders. Its members are more numerous, more varied, and more widely distributed than those of any other suborder. Some are among the most terrestrial of all salamanders, but the newts
and *Amphiuma* are highly aquatic. Because they are so varied, it is difficult
to define the suborder in a few words. Males usually have three sets of
cloacal glands, spermatophores are produced, and fertilization is always
internal. In the palatal region of the skull, a dentigerous (tooth-bearing)
process from the prevomer extends back along the side of the parasphenoid.
Sometimes the tooth patches are split off and lie on top of the parasphenoid.
Any metamorphosed salamander that has teeth on the roof of the mouth
well behind the internal nares belongs in this suborder. Salamandroids are
found in Europe, Asia, and North America and a few reach into northern
South America and northern Africa. They are divided into three families.

Family Salamandridae. These are the typical salamanders and newts.
They usually metamorphose completely and spend at least part of their lives
on land, though occasional neotenic ones are found. A salamandrid has
lungs and an ypsiloid cartilage. The prevomerine teeth are in two long rows,
one on either side of the parasphenoid. In ancient times, the European
Salamandra was credited with being able to live in fire. It is easy to see how
such a myth arose. Salamanders sometimes take shelter in crevices in fallen
logs. When such a log is thrown on a fire, the animal is roused by the heat
and tries to escape, seeming to the unknowing to emerge from the flame.
Even today, objects that can withstand a great amount of heat are sometimes called salamanders.

Considerable confusion often arises about the words salamander, newt,
and eft. *Salamandra* is a Greek word meaning a lizardlike animal. It is used
as the generic name for the common terrestrial tailed amphibians of Europe.
As a common noun the word salamander is applied generally to all the
caudate amphibians, and especially to the more terrestrial salamandroids.
Newt and eft come from the Anglo-Saxon *efete* or *evete*, a word used for
both lizards and salamanders. In medieval English this word became ewt
and finally a newt (an ewt). Since the only caudates found in England are
moderately aquatic members of the family Salamandridae, newt has come
to be used as the common name of the more aquatic salamandrids. Larvae of
some of the American newts (*Diemictylus*) frequently metamorphose into
tiny spotted creatures, often bright red or orange in color, that leave the
water and live on land for as much as three years. This land stage is called

an eft, red eft, or spring lizard. The eft later returns to the water, develops a tail fin and adult coloration, and changes into a sexually mature, aquatic newt. This condition is not fixed, even in a single species, for the larva may metamorphose directly into the aquatic adult. The eft stage is not known in European newts, but the adults frequently spend part of their time on land.

Salamandridae is a widespread family, occupying Europe, eastern Asia, and both sides of North America. It includes about 40 species, most of them Palearctic forms. Best known are members of the genus *Salamandra* and the European newts, *Triturus*. *Pleurodeles*, *Triturus*, and *Salamandra* extend into northern Africa and are the only salamanders known from that continent. Only two genera of newts, *Diemictylus* in the east and *Taricha* in the west, occur in North America.

The salamandrids exhibit a wide variety of rather elaborate courtship patterns. Sight, smell, and touch may all play a part in arousing the female to pick up the spermatophore. During the breeding season, males of some species develop vivid colors and special structures, such as high dorsal crests, which they display before the females. The male may also stimulate the female by the secretions of special hedonic glands, by rubbing, prodding, nipping, or carrying her on a "piggy-back" ride. The male Mountain Newt of Europe (*Euproctus*) actually clasps the female and may place the spermatophore directly in her cloaca. Variation in courtship pattern is probably an important isolating mechanism.

The Red-spotted Newt of eastern North America (*Diemictylus v. viridescens*) mates in water in the early spring. The male seizes the female and rubs her with his cheeks and chin, thus smearing her with the odorous secretion of his hedonic glands. He then moves off a short distance and deposits a spermatophore. She follows, takes it into her cloaca, and some hours later, begins to lay. A female may deposit from 200 to 375 eggs, sometimes over a period of several months. The eggs are laid singly, usually fastened to a leaf or the stem of a small plant, in quiet waters; less often they may be attached to the surface of a stone. The egg is pigmented, with the animal pole varying from light to dark brown. The period of incubation varies from about 20 to 35 days. The larva at hatching is about 7 or 8 mm. long and has well-developed balancers, one on each side of the head just below the eye. The front legs are short, blunt buds and the hind legs are undeveloped. As in the larvae of ambystomatids and most other salamanders that breed in quiet waters, a well-developed dorsal keel, continuous with the tail keel, extends nearly to the base of the head. Metamorphosis usually takes place in late summer or early fall, after a larval period of two or three months. The European *Salamandra salamandra* and *S. atra* retain the eggs in the oviducts for at least a part of the developmental period. The young of *S. atra* are born as fully metamorphosed individuals. The female *S. salamandra* goes to water to bear the young which are usually born as late larvae. If the embryos of these two species are dissected from the oviduct, they are found to have long, filamentous gills and rudimentary balancers, indicating that these forms evolved from salamanders having pond-type larvae.

Family **Amphiumidae.** This is the smallest of the families of salamanders. The only living form is *Amphiuma means,* the aquatic Congo Eel of the southeastern United States. *Amphiuma* is a big, dark-colored, semilarval animal, with a long, cylindrical body and tiny, useless arms and legs. Large specimens may be more than 90 cm. long. Adults lose the gills though one pair of gill slits remains open. They also show other larval characters, such as the lack of eyelids. They have lungs but no ypsiloid cartilage. These salamanders are savage, and large ones can inflict a painful bite.

Much remains to be discovered about the breeding habits of *Amphiuma.* Fertilization is apparently internal. Even though the adults are primarily aquatic, the eggs are laid in shallow depressions on land, beneath old logs or boards, and are guarded by the female. The rosary-like strings contain about 150 eggs, each about 9 mm. in diameter. The newly hatched larvae are from about 60 to 75 mm. in total length, of which about 10 mm. is tail. They have short, white gills, and are dark brown above and on the sides. Many small, round, light dots are scattered over the surface.

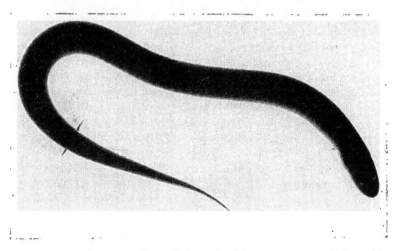

FIG. 12-5. The two-toed Amphiuma, *Amphiuma m. means,* of the southeastern United States. Its eel-like form has led to the common name of Congo Eel.

Family Plethodontidae. This is the largest and most successful group of living salamanders. Its members are mostly small to medium in size, and, for salamanders, they occupy a wide variety of habitats. Some are aquatic, some terrestrial, some are burrowers, and a few are arboreal. *Typhlomolge, Haideotriton,* and *Typhlotriton* are all blind, white salamanders found in caves and underground waters. The first two are permanent larvae, but *Typhlotriton* metamorphoses and only the adults are blind. All plethodontids are lungless and lack the ypsiloid cartilage. A small, gland-lined groove, the nasolabial groove, runs from the nostril to the upper lip on each side. It apparently helps keep the nostril clean and free of water. It may also transfer olfactory sense data to the nose. The Plethodontidae are concentrated in North America, but one group (*Bolitoglossa* and related forms) extends into northern South America and another genus (*Hydromantes*) has species

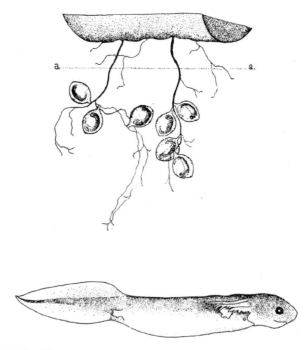

FIG. 12-6. The eggs and newly hatched larvae of the Rusty Mud Salamander, *Pseudotriton montanus floridanus;* a———a represents the water line. [After Goin.]

in California and southern Europe. Most of the common land salamanders of the United States belong in this family and it is the one being most actively studied in this country today.

Life histories of the Plethodontidae include all gradations between the aquatic and the completely terrestrial. Hedonic glands are well developed and widely distributed over head, body, and tail of most male plethodontids. Courtship includes rubbing and prodding of the female, and the "tail-walk" in which the pair moves along with the female straddling the tail of the male. *Pseudotriton montanus floridanus* is an aquatic type. The eggs are deposited in small groups on tiny rootlets and other submerged objects in cool, muddy springs. In a single clutch, found hanging from rootlets at the edge of an undercut bank, there were 27 eggs in tiny clusters of from 2 to 8. The female stays with the eggs. How long embryonic development takes is not known, but the young hatch at lengths of about 12 to 14 mm. Balancers are absent in this species.

The Dusky Salamander, *Desmognathus fuscus,* lays its eggs on land rather than in water. They are deposited in small clusters, in shallow excavations in soft earth, within beds of sphagnum, or beneath stones or logs. These excavations, or nests, are generally located from 30 to 100 cm. from the water. Five egg clusters of the southern race contained 9, 11, 14, 15, and 19 eggs. These eggs were approximately 5 by 7 mm. in size and each was attached by a short, twisted stalk, about 2 mm. in length, to a common base along a rootlet. The whole cluster had the appearance of a bunch of grapes. When the young hatch they are from 16 to 20 mm. in total length. They do not go to the water at once, but remain in the nest with the mother and show definite terrestrial adaptations. The hind limbs are longer in proportion to the trunk region than at any time during later development. These young salamanders are thus not merely little larvae that have not had a chance to reach the water, but are basically terrestrial salamanders, able to move about in the damp crannies and crevices leading from the nest to the nearest pool or stream. After one or two weeks, they enter the water to exist as aquatic larvae until they are 7 to 9 months old. They are about 45 mm. in length at the time of metamorphosis.

The Red-backed Salamander, *Plethodon cinereus,* of the eastern United States, has a typical terrestrial plethodontid life history. The females lay 3 to 12 large, unpigmented eggs in crannies and holes in partly decayed logs. Each egg adheres to those laid before to form a little mass of eggs, seemingly contained in a single envelope. This egg cluster is usually attached to the roof of a small cavity in the log. The embryos develop rapidly and soon show large, well-developed, external gills. These are lost on hatching and the

FIG. 12-7. The Slimy Salamander, *Plethodon glutinosus*, a typical plethodontid salamander.

young have the same form as the adults. They never take up an aquatic larval existence.

Suborder Proteida

This suborder is made up of aquatic salamanders, permanent larvae that retain the gills and two pairs of gill openings as adults. They develop lungs, but lack nasolabial grooves, an ypsiloid cartilage, maxillary bones, and eyelids. The angular and prearticular are fused. The presence of so many larval characters makes it difficult to determine the relationship of this suborder to any of the other primary groups of salamanders. It seems to be a distinctive, small natural group of perennibranchs. There is only one family in the suborder.

Family Proteidae. There are only two genera. *Proteus*, the blind, white cave salamander of Europe, has a long, pigmentless body, bright red gills, and skinny appendages provided with only three fingers and two toes. Members of the genus *Necturus* have pigmented bodies, stouter limbs, and four toes on each foot. These are the mudpuppies of the eastern United States. There are four species of *Necturus*, and only one of *Proteus*.

Fertilization is internal in the proteids. Little is known of the actual courtship procedure. In *Necturus maculosus*, mating apparently takes place in the fall and the eggs are laid the following spring. The nests are excavations beneath stones, boards, or other objects, lying in water at depths of 10 to 150 cm. In a lake habitat, nests are found at distances of 4 to 8 m. from shore. They have also been recorded from streams. The eggs are deposited singly and are attached to the undersurface of the object sheltering the nest. Clutch size varies from 18 to 180, with a tendency for clutches deposited in streams to be larger. For example, in five lake nests the average number of eggs per nest was 66, whereas in three stream nests the average number was

107. The eggs, which may be as large as 10 or 11 mm. in diameter, hatch after about 4 or 5 weeks. The newly hatched larvae range from 20 to 25 mm. in length. There is, of course, no metamorphosis, since these salamanders are perennibranchs.

Proteus usually lays eggs but sometimes retains them in the oviduct where one or two of the young undergo development. They are born as miniature replicas of the adult. No special modifications, either of the larvae or of the oviducts, are known to accompany this change in life history.

Collateral Reading and General Reference

Angel, F. *Vie et Moeurs des Amphibiens.* Paris: Payot, 1947.

Bishop, S. C. *Handbook of Salamanders.* Ithaca: Comstock, 1947. (The single, modern, comprehensive account of North American salamanders.)

Boulenger, G. A. *Catalogue of the Batrachia Gradienta s. Caudata and Batrachia Apoda in the Collection of the British Museum.* 2d ed. London: British Museum, 1882. (The last comprehensive report of the salamanders of the world, now rather out of date.)

Cope, E. D. "Batrachia of North America." *Bulletin of the United States National Museum,* no. 34, 1889. (The classic in its field. Although the nomenclature is somewhat out of date, this is still a fundamental work on the classification of amphibians.)

Dunn, E. R. *Salamanders of the Family Plethodontidae.* Northampton: Smith College, 1926. (Another classic. The preface and introduction should be read by all students of herpetology.)

"American Caecilians." *Bulletin of the Museum of Comparative Zoology at Harvard University,* Vol. 41, no. 6, 1942. (The only comprehensive work on New World Apoda.)

Noble, G. K. *Biology of the Amphibia.* New York: McGraw-Hill, 1931.

Stebbins, R. C. *Amphibians of Western North America.* Berkeley and Los Angeles: University of California Press, 1951. (A thorough and excellent account of a rather large fauna within a limited area.)

FROGS AND TOADS

ALTHOUGH THE FROGS AND TOADS are seldom of much interest to amateurs, for a number of very practical reasons they (particularly members of the genus *Rana*) are of prime importance to professional zoologists. Their tendency to congregate in large breeding choruses makes it simple to collect large numbers of them. They are easy to maintain alive or to preserve, large enough to be worked on conveniently, small enough not to raise serious problems of storage space. Cheap and convenient, they are the laboratory animals *par excellence* of the introductory biology course. Because they lay large numbers of eggs that are easy to collect and to maintain and hatch in the laboratory, because these eggs are surrounded by a clear jelly rather than an opaque shell, and because most frogs pass through a free-swimming larval stage before developing the adult form, they are ideal for studies in experimental embryology. They have further potentialities as research animals that are only now beginning to be explored. In the field of ecology, studies of social hierarchies among animals have recently been extended to include the African Clawed Frog, *Xenopus*. Some of the tropical genera speciate readily and have developed hundreds of distinct forms. These genera may give us clues to mechanisms of speciation that have so far eluded us in studies of more restricted genera.

All frogs are included in the order Anura. They are easily recognized, and there is never any doubt whether a particular animal is a member of this order. Each frog lacks a tail in the adult stage and has the head and trunk fused so that there is no neck. It has well-developed legs, the hind ones being longer than the front ones. Since the great majority of the more primitive frogs are aquatic, it seems logical that the long, muscular hind legs developed first as swimming organs and only later became specialized for jumping. Frogs are the lowest form of vertebrates to have a middle ear

216

cavity. The tympanic membrane usually lies flush with the surface of the head and is often a very large and obvious structure. Correlated with the development of the ear as a hearing organ is the appearance of a true voice box. Movable eyelids protect the eyes and glands are present that keep them moist.

Although the modern frogs and toads are remarkably uniform in gross structure, they have undergone considerable adaptive radiation in their reproductive habits. Fertilization is almost always external. Usually the eggs are laid in water and hatch into aquatic larvae. But all stages are found between this reproductive pattern and true terrestrialism in which the eggs are laid on land and the young hatch as tiny frogs.

Most anurans require moist surroundings and cannot stand prolonged exposure to low humidities, but a few, such as the Giant Toad, *Bufo marinus*, are able to adapt to rather arid conditions. The group as a whole is well adapted to a relatively few niches and in those niches has spread round the world. With the possible exception of the lizards, they are the most widely distributed of the herptiles.

Attempts to divide the Anura into suborders and families have met with many obstacles. As a group, the frogs show a remarkable tendency toward parallelism and convergent evolution. Many characters that were once thought to be indicative of relationship have been shown to have evolved independently in several different lines. No single feature can be relied on to separate the frogs into natural groups. We must consider a complex of characters, reproductive as well as morphological. Unfortunately, relatively few species have been thoroughly studied anatomically, and the life histories of many are completely unknown. The accumulation of further knowledge will undoubtedly require modifications in the classification given below, and it should in no sense be considered final.

One of the most important characters used in the classification of the anurans is the shape of the centrum. In the ontogenetic development of the vertebrae, the centra and intervertebral bodies form as, respectively, thin and thick cylindrical regions alternating along the length of the perichordal tube (see Fig. 13-1). The intervertebral elements chondrify more rapidly than the centra and expand inward to constrict the notochordal tube. Later the centra ossify so that the vertebral column is made up of a chain of ossified centra joined by undivided intervertebral cartilages. Each intervertebral body is then invaded by an arc of connective tissue. Later a split develops along this arc, dividing the cartilage into two unequal parts, a small cup and a larger ball, which remain attached to their respective centra. Thus the anterior portion forms the posterior face of the centrum in front and the posterior portion forms the anterior face of the centrum behind. Which

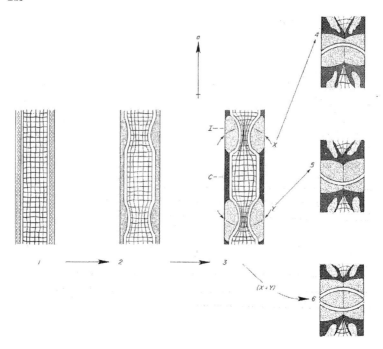

FIG. 13-1. Diagrammatic representation of vertebral development in Anura. 1. Early larval stage; 2. 22 mm. tadpole of *Rana temporaria*; 3. *Rana temporaria* 2 years of age; 4 and 5, respectively, opisthocoelous and procoelous vertebrae of *Rana temporaria*, 3–4 years old; 6, "free disc" condition of *Megophrys major*, 4 years of age. X and Y = invading connective tissue arcs; (X + Y) = invasion of a single intervertebral body by two connective tissue arcs. C, ossified cylinder of centrum; I, cartilaginous intervertebral disc. [After Griffiths.]

portion forms the cup and which the ball is determined by the direction of the slope of the arc of invading connective tissue. If the slope runs antero-posteriorly, the ball is formed from the anterior part of the intervertebral body, and the cup from the posterior part. This results in a vertebra with a centrum hollowed out in front and rounded posteriorly—a procoelous vertebra. If the slope runs the other way, the vertebral centrum is rounded anteriorly and hollowed posteriorly—an opisthocoelous vertebra. If the direction of slope alternates in successive intervertebral bodies, a vertebra is formed that has the centrum hollowed at both ends—an amphicoelous vertebra. When the presacral vertebra is amphicoelous and the other vertebrae are procoelous, the vertebral column is said to be diplasiocoelous;

that is, it has two types of vertebrae. Sometimes the intervertebral body is invaded by two arcs of connective tissue sloping in opposite directions. This results in the formation of a free intervertebral disc between the centra of successive vertebrae.

Another character that has been much used in the classification of frogs is the construction of the pectoral girdle. If the epicoracoid cartilages on the ventral ends of the two halves of the girdle are separate, with one over-lapping the other, the girdle is said to be arciferal (arch-bearing, from the shape of the epicoracoids). If these cartilages are fused together, the girdle is said to be firmisternal. However, not all frogs have girdles that can be classified as either arciferal or firmisternal. In some (e.g., the Cuban *Sminthillus*) the epicoracoids may be fused anteriorly, but separate and overlapping posteriorly. Others (e.g., *Sooglossus* of the Seychelles Islands and some species of *Rana*) have the epicoracoids fused anteriorly, then overlapping, then again fused posteriorly. Furthermore, it was long believed that the arciferal girdle was the primitive one from which the firmisternal girdle evolved. But the oldest known frog, the Jurassic *Notobatrachus*, apparently had a firmisternal girdle. It may be, then, that this is the truly primitive condition, from which varying degrees of epicoracoid freedom have evolved in several different lines (see Fig. 13-2).

SUBORDER AMPHICOELA

These are the most primitive frogs living today. The name of the suborder is unfortunate, since the vertebrae are not truly amphicoelous. Instead, the intervertebral bodies remain undivided throughout life so that the centra are joined to each other by cartilaginous rings. There are nine presacral vertebrae. Like all frogs, the members of this suborder lack tails, but they show their descent from tailed ancestors by the presence of two tiny "tail-wagging" muscles with formidable names—pyriformis and caudalipubo-ischiotibialis. Free ribs (not fused to the vertebrae) are present in the adults. The girdle is arciferal in living species. The suborder has only one living family.

Family Leiopelmidae

These are small grayish frogs, sometimes daintily patterned with pink, brown, and yellow. There are only two genera—*Leiopelma* in New Zealand and *Ascaphus* in northwestern United States. This widely disjunct distribution is an indication of the antiquity of the group. *Ascaphus* is unique among

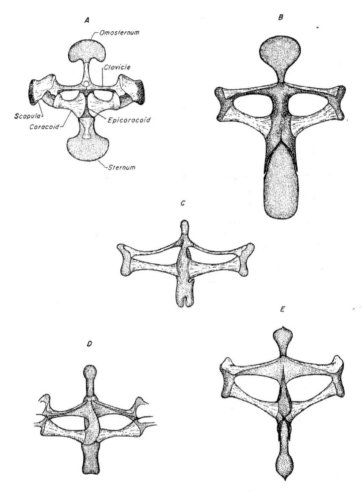

FIG. 13-2. Pectoral girdle types in frogs. A, The firmisternal girdle in *Rana tigrina*; and variations in the arciferal type as seen in: B, *Rhinoderma darwini*; C, *Sooglossus seychellensis*; D, *Eleutherodactylus bransfordi*; and E, *Sminthillus limbatus*.

frogs in that an extension of the cloaca of the male, the so-called "tail," is used as an intromittent organ for the transmission of sperm to the cloaca of the female.

Ascaphus breeds in swift-flowing mountain streams. The male is voiceless.

FIG. 13-3. The Tailed Frog, *Ascaphus truei*, of the Pacific coast of North America. The tail is really the everted cloaca which is used as an intromittent organ.

Instead of sitting still and calling for a mate, he swims about on the bottom until he finds one. He clasps her with an inguinal embrace, humps his body, and maneuvers his extended cloacal appendage into position to thrust into her cloaca. The intromittent transfer prevents the sperm from being swept downstream before they can enter the ova. The eggs are deposited in coils of rosary-like strings which adhere to rocks at the bottom of the stream. In the cold water in which these eggs are deposited, embryonic development is slow and transformation does not occur until the following summer.

Leiopelma lays its eggs in damp places on land. The young hatch into terrestrial froglets, though they still have well-developed tails which are resorbed during the first month of growth. The places in which the eggs are deposited are subject to flooding; if the eggs are washed into the water, they are still able to complete normal development. Furthermore, if the embryos are prematurely released from the egg capsules, they can be reared as tadpoles in water.

SUBORDER AGLOSSA

As the name indicates, the members of this suborder lack a definitive tongue. The vertebrae are opisthocoelous. The urostyle is either fused to the sacrum or articulated to it by a single condyle. The shoulder girdle is partly or wholly firmisternal and the ribs, though free in the tadpole, are fused to the vertebrae in the adult. The suborder includes only one living family.

Family Pipidae

These are stout-bodied, big-footed, highly aquatic toads. Some forms have the surface of the skin covered with horny tubercles and lateral line sense organs are present in some adults. The Pipidae have the disjunct distribution so common in ancient and primitive stocks. They are found in Africa south of the Sahara and in South America.

The African pipids *Xenopus, Hymenochirus,* and *Pseudhymenochirus* have the tips of the digits undivided and the sternum large. *Xenopus,* the Clawed Frog now so often used in pregnancy tests, is the most primitive and has the widest distribution. All of the South American species are now placed in the single genus, *Pipa.* The tips of the fingers of these frogs are divided to form starlike structures, and the sternum is small.

More striking than the differences in structure that distinguish the forms of the two continents are the differences in life history. The African pipids deposit their eggs in water. The male, at the time of mating, makes a sudden dash at the female and clasps her by the inguinal region. The female ovulates each egg separately, holding it for a moment by her cloacal lips. She then grasps a leaf or twig with her hind limbs. The egg is ejected sharply by the cloaca and is propelled along a shallow groove on the venter of the male, past his cloaca and then to the weed, to which it becomes attached.

The eggs of members of the genus *Pipa* develop in temporary, individual pits formed in the soft skin on the back of the female. The method by which the eggs are transferred to the back of the female has been described for a pair of Surinam Toads (*Pipa pipa*) that bred in an aquarium. The male clasped the female inguinally as the pair rested on the bottom of the tank. The dorsal part of the female's vent was pressed against the abdomen of the male. The pair then rose to the top, turning over as they did so, and paused momentarily in an upside down position. At this point three to five eggs were extruded by the female and caught in transverse skin folds on the belly of the male. The two then returned head first to a tilted resting position on the bottom. Fertilization apparently occurred during the descent, and as the frogs righted themselves, the eggs dropped to the back of the female and adhered there. A number of turnovers were required to complete deposition. Fifty-five eggs were thus implanted on the back of the female, and eleven more were lost as they were laid. Eggs of some species of *Pipa* hatch into tadpoles, but others hatch directly into tiny frogs.

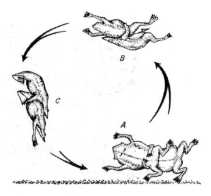

SUBORDER OPISTHOCOELA

Members of this suborder have arciferal girdles and opisthocoelous vertebrae. The urostyle is not fused

FIG. 13-4.
The egg-laying turnover maneuver of breeding Surinam Toads, *Pipa pipa*. The eggs are extruded by the female when both individuals are upside down, B. [After Rabb and Rabb.]

to the sacral vertebra, but articulates with it by one or two condyles. There are eight presacral vertebrae. The suborder includes two recent families.

Family Discoglossidae

Adults of this family of rather primitive frogs have free ribs, a tongue, and movable eyelids. An anterior extension of the procoracoid cartilage splits off to form a separate element, the omosternum. The four genera are all Old World forms. Members of the genus *Bombina*, found from Europe to China, are rather flattened, highly aquatic toads. They are brilliantly marked below, black mottled with bright red, orange, yellow, or white. *Discoglossus* is a rana-like frog found in southwestern Europe and northwestern Africa. *Alytes* of western Europe, smaller and more terrestrial, includes the famous Midwife Toad. The completely aquatic *Barbourula* is known only from the Philippines.

Bombina maxima, the Yellow-bellied Toad, breeds in water in the spring and summer. The male clasps the female with a pelvic embrace. The eggs are laid in small masses which sink to the bottom, or come to rest suspended on vegetation, and hatch in about a week.

FIG. 13-5. The Korean Fire-bellied Toad, *Bombina orientalis*.

The breeding habits of *Bombina* are probably fairly typical for most other members of the family, but the genus *Alytes* shows one of the most remarkable modifications of any frog. The male of the Midwife Toad, *Alytes obstetricans,* calls from a small hole in the ground. Mating takes place on the ground and is apt to last most of the night. The male clasps the female tightly around the head, above the forelimbs. Just before ovulation, he moves his hind legs forward so that his heels are together, anterior to and above the cloaca of the female. As the eggs are emitted, he catches the rosary-like strings in his feet and, by stretching his legs backward, delivers from 20 to 60 eggs. He then moves his legs so as to twist the strings around them. He carries the eggs thus for several weeks, until the tadpoles are about ready to hatch. Then he makes a brief visit to a pool where no other tadpoles are present. Here the eggs hatch and the tadpoles finish their development.

Family Rhinophrynidae

The Mexican Burrowing Toad (*Rhinophrynus*), the only member of this family, is an egg-shaped, short-legged, toothless creature with a smooth,

blotched skin. This aberrant and highly specialized frog is difficult to classify, but is perhaps more closely allied to the discoglossids than to any
other group. The omosternum is rudimentary and the sternum absent. Free
ribs are not present at any stage. The tongue is free, rather than attached
in front, and the toad is apparently able to protrude it like a mammal
rather than to flick it out in normal toad fashion. The foot is specialized
for digging—the prehallux is covered with an enormous, cornified "spade"
and the single phalanx of the first toe is shovellike. *Rhinophrynus* apparently feeds largely on termites. Amplexus is inguinal, the eggs are deposited in water, and the tadpoles are free-swimming.

SUBORDER ANOMOCOELA

The pelobatid toads and their allies lack ribs at any stage and have true
teeth in the upper jaw. The shoulder girdle is arciferal. An omosternum
is present. The vertebrae are either procoelous or amphicoelous with free
intervertebral discs. The sacral diapophyses (the vertebral processes to
which the pelvic girdle is attached) are expanded. The suborder includes
only two families.

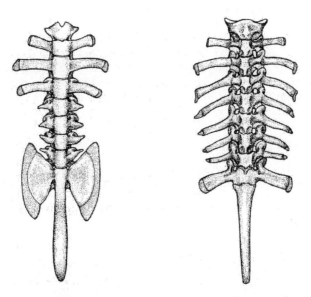

FIG. 13-6. Vertebral columns of (left) a frog with expanded sacral diapophyses, *Scaphiopus couchi;* and (right) one with rounded sacral diapophyses,
Rana virgatipes. [*After Noble.*]

Family Pelobatidae

In the pelobatids the sacral vertebra is either fused to the urostyle or articulates with it by a single condyle. The astragalus and calcaneum, two long bones in the ankle, are separate.

Subfamily Megophryinae. The Megophryinae have free intervertebral discs. The subfamily includes about half a dozen genera and many species, some of them quite large and spectacular. They are found in southeastern Asia, from China to the East Indies. The quaint Nose-horned Frog of the East Indies (*Megophrys nasuta*) has each upper eyelid extended to form a large, thin, pointed "horn," and a flexible projection from the snout, so that the animal looks as though it had three horns.

Subfamily Pelobatinae. These are pelobatids with procoelous vertebrae. The Spadefoot Toads of Europe and northern Africa (*Pelobates*) and of North America (*Scaphiopus*) are members of this subfamily. The "spade" is a broad, sharp-edged tubercle on the inner side of the hind foot, with which these burrowing toads can dig themselves out of sight in short order. Spadefoots resemble the true toads (*Bufo*) in build and are often confused with them. They may be distinguished by their smoother, less warty skins, and vertical, rather than horizontal, pupils.

Pelobatids usually lay their eggs in open water and pass through a tadpole stage. *Scaphiopus* lays in temporary ponds and eggs and young sometimes show a remarkably rapid rate of development. The eggs may hatch in a day and a half and the tadpoles may transform in fifteen days. *Megophrys longipes* of Malaya is said to lay about a dozen large eggs in humid moss with the young metamorphosing before hatching.

Family Pelodytidae

This family includes only one living genus, the slender *Pelodytes* of Europe. The sacrum is free from the urostyle, which has a double condyle. The astragalus and calcaneum are fused into a single bone. Reproduction is aquatic.

SUBORDER DIPLASIOCOELA

The suborders discussed above are primitive, limited in distribution, and represented today by relatively few species. The suborder Diplasiocoela is an extensive, widespread, successful group, comprising a variety of highly diverse forms. The shoulder girdle is usually firmisternal. Sometimes the epicoracoids are more or less separated, but they are never produced into posteriorly directed epicoracoid horns, as they are in the suborder described next. The vertebral column is either diplasiocoelous or uniformly procoelous. The sacral vertebra articulates with the urostyle by a double condyle (except in the Sooglossinae). Ribs are never present. The suborder includes four families.

Family Ranidae

These are the so-called "true frogs." They are distinguished from the other diplasiocoels by their cylindrical or very slightly dilated sacral diapophyses in conjunction with the absence of an extra element, the intercalary cartilage, between the next to the last and last phalanx of each digit. Africa seems to have been the center of differentiation of the ranids. Six of the eight subfamilies here recognized are found there, and four of them are found nowhere else. One subfamily, Dendrobatinae, occurs only in tropical America. The genus *Rana*, of the subfamily Raninae, is almost cosmopolitan in distribution.

Subfamily Arthroleptinae. These little African ranids have horizontal pupils and lack vomerine teeth. A few have uniformly procoelous vertebrae. The sternum may be either cartilaginous or bony. (It is usually cartilaginous in the small individuals and species and bony in the larger ones.) Males of a few species of *Arthroleptis*, and of the genera *Schoutedenella* and *Cardioglossa* have the third finger greatly enlarged. It is sometimes as much as two or three times as long as the rest of the hand—a striking secondary sexual character unknown among other amphibians. The subfamily includes about six genera and many species; all are little frogs of the forest floor or brushy country.

Subfamily Sooglossinae. This subfamily includes two genera, *Sooglossus* and *Nesomantis*, both restricted to the Seychelles Islands northeast of Madagascar. The intervertebral body between the sacral vertebra and the urostyle remains undivided. The epicoracoid cartilages are fused anteriorly and posteriorly, and overlap mesially. *Sooglossus* resembles a small arthroleptid externally, whereas *Nesomantis* is more toadlike in build.

Subfamily Dendrobatinae. This is another small subfamily, comprising three genera, *Dendrobates*, *Phyllobates*, and *Prostherapis*. They are all smooth-skinned little frogs, found only in tropical America. The vertebral column is uniformly procoelous. (In *Dendrobates* the sacral and presacral vertebrae are fused.) The girdle is firmisternal, the omosternum frequently

FIG. 13-7. The Arrow Poison Frog, *Dendrobates auratus*, of Central America.

bony. A pair of glandulo-muscular organs is present on the upper surface of the tip of each digit.

Subfamily Astylosterninae. These are West African forest frogs of average size, characterized by having vertical pupils and a forked, bony omosternum. The subfamily seems to be a natural group of four closely related genera. The terminal phalanges of both fingers and toes of *Nyctibates* are simple and only slightly curved; those of two or more toes of *Scotobleps*, *Astylosternus*, and *Gampsosteonyx* are bent sharply downward and may pierce the integument to form bony little spines which protrude below the tips of the digits. The functional significance of these modified toes is not known. It may be that they give a sure grip for jumping. The genus *Astylosternus* includes the famous Hairy Frog, *A. robustus*, a species in which the male has a peculiar growth of hairlike processes on the thighs and flanks. These are most fully developed during the breeding season. They are not true hairs, but vascular papillae that presumably aid in respiration.

Subfamily Phrynopsinae. These small, rana-like African frogs have horizontal pupils, vomerine teeth, and a cartilaginous unforked omosternum. *Phrynopsis* of Cameroun and Mozambique is big-headed, with elongate, spiky teeth. *Leptodactylodon* of Cameroun is small-headed and has slightly dilated digital discs.

Subfamily Raninae. These ranids have a bony sternum and pointed, or slightly dilated, digit tips that lack discs on either the upper or lower surfaces. This subfamily includes the genus *Rana* and its close allies. About 200 species of *Rana* have been described. Of the major land masses, only Greenland, New Zealand, central and southern Australia, and southern

FIG. 13-8. A New Jersey specimen of the Bullfrog of the eastern United States, *Rana catesbeiana*. This frog has now been introduced into various regions of the world as a potential food supply.

South America lack at least one representative of this vigorous genus. Many are large, and most are colored with soft, muted browns, greens, and yellows. Most of the other genera contain only a few forms and have a limited distribution in Africa or southern Asia.

The largest known frog, *Conrana goliath* of Africa, belongs to this subfamily. Some say that this giant of all frogs is really nothing but a large *Rana*, but it differs from that genus in the weak calcification of its epicoracoid cartilages.

Subfamily Petropedetinae. This small group of African ranids includes two genera, *Petropedetes* and *Arthroleptides*. Both have a pair of glandulomuscular organs on the upper surface of each digit. In this respect, they are like the South American dendrobatines. The omosternum is bony and either entire or slightly forked posteriorly. The terminal phalanges are T-shaped.

Subfamily Cornuferinae. These are ranids that have the tips of the digits more or less dilated and showing slight indications of the friction pads characteristic of the tree frogs of the following family. The group extends from Africa across the Indo-Australian archipelago to the Solomon and Fiji Islands, and from China to northern Australia. A great many genera and species have been described, but many of them tend to grade into one another and are difficult to define. The subfamily as a whole, though, is a unified group and is apparently the one that gave rise to the family described next.

Very little is known of the reproductive habits of many of the Ranidae, particularly of the African forms. The "typical life history of a frog," as is described in most biology books, is almost invariably based upon some species of *Rana*. Most of the species of this genus for which life histories are known conform to the pattern of laying a large number of small-yolked eggs in open water and having a free-swimming tadpole stage. A few African and Oriental species lay their eggs out of water on leaves or stones or in mud burrows, but the egg masses are unmodified and after hatching the tadpoles soon make their way into the water. *Staurois* (subfamily Cornuferinae) is a mountain torrent form, with many species in southeastern Asia. The tadpole has a distinctive sucking disc back of the mouth for clinging to rocks in the swift water in which it lives. This abdominal disc is not present in the tadpole of any other frog. Members of the genus *Cornufer* of the East Indies lay large-yolked eggs on land and development is direct. At least some species of the African *Arthroleptis* also lay eggs on land.

Males of the subfamilies Sooglossinae and Dendrobatinae transport the tadpoles on their backs from the egg-laying site on land to water. The female of *Dendrobates auratus* lays from one to six large-yolked eggs surrounded by an irregular, sticky, gelatinous material with no definite external film. The male either guards or periodically visits the clutch. The eggs hatch in about two weeks and the newly hatched tadpoles wriggle onto the back of the male. He then visits water, the young leave his back, and finish their development as free-swimming tadpoles. They may not transform until six weeks later. Males carrying tadpoles have been found in trees some distance from any pond or stream, but they may have been about to deposit their offspring in tree holes containing water. Tadpole-carrying males have also been reported in the genera *Phyllobates* and *Prostherapis*. A specimen of *Prostherapis fuliginosus* with 25 tadpoles on his back has been taken. The larvae of *Sooglossus seychellensis* lack gills and hatch with the hind leg rudiments already developed.

Family Rhacophoridae

The dozen-odd genera of this family are diplasiocoelous (rarely procoelous) frogs with cylindrical sacral diapophyses. An intercalary cartilage is present between the penultimate and ultimate phalanx of each digit. The extra joint thus provided allows the last phalanx, with its adhesive disc, to be placed flat against the surface regardless of the position of the rest of the foot—an obvious advantage to a climbing form. Most rhacophorids are tree frogs, many of them large-disced, brightly colored forms. A few have given up the arboreal habit and returned to the ground, but they retain the adhesive mechanism and intercalary cartilages of their relatives. The family inhabits southern Asia, Japan, the Philippines, the East Indies, Africa, and Madagascar.

Rhacophorids typically lay their eggs in masses of foam on the leaves of plants or other structures above water. The female does most of the work in producing the foam mass. Before the eggs appear, she ejects a small amount of fluid which she beats into a froth by moving her feet mesially and laterally and turning them as she crosses them on the midline. When the foam has been prepared, the eggs and fluid are ejected together. During the egg-laying process, the male is passive, grasping the female under her arm pits and simply holding his body closely applied to her back, his eyes half closed. His pelvic region is bent down with the cloacal opening near that of the female and the eggs are apparently fertilized as they leave the cloaca of the female. When the egg-laying process is concluded, the female stands up on her forelimbs and the male tries to get away from the foam in which the distal ends of his hind legs are buried. The female

usually disentangles herself later by moving her legs and body sideways
with the help of large, sticky, finger discs.

After the egg-laying, the foam changes color from white to light brown.
The eggs are scattered singly or in small groups in the foam mass but are
mostly concentrated near the basal part where the foam is attached to the
substrate. When the time of hatching is near, the foam begins to liquefy
and the active movement of the fully developed embryos or tadpoles drops
them into the water below. Sometimes the whole egg foam mass may be
washed down by rain into the pool. Tadpoles of a few species go through
their entire development in the nest.

African rhacophorids of the genera *Hyperolius* and *Kassina* lay small,
pigmented eggs directly in the water without the foam nest. In contrast,
Rhacophorus microtympanum of Ceylon lays about 20 large-yolked eggs
on land. The female does not produce a foam nest, but remains with the
clutch. Development is direct. The embryos bear a striking resemblance to
embryos of *Eleutherodactylus,* a member of another suborder described
below, which also develop directly on land.

Family Microhylidae

These are truly members of the suborder Diplasiocoela, although many
of them have a vertebral column that is procoelous. They show their relationship to the other diplasiocoels in the disposition of their thigh musculature (the semitendinosus remains distinct from the sartorius, its distal
tendon passing dorsal to the distal tendon of the gracilis mass). They are
firmisternal, but differ from the ranids in having the sacral diapophyses
more or less dilated and from the rhacophorids in lacking the intercalary
cartilages. Teeth are frequently lacking and the pectoral girdle is reduced.
For the most part, they are dull-colored, inconspicuous little frogs. The
family apparently originated in southeastern Asia and has spread from there
to New Guinea, Africa, including Madagascar, and to the Americas. It has
been divided into seven subfamilies.

Subfamily Dyscophinae. These are microhylids that retain the maxillary
and vomerine teeth and have large and undivided prevomers on the roof
of the mouth. Two genera are found in the Indo-Malayan region and one
in Madagascar. The eggs are laid in water and the tadpole is a characteristic microhylid tadpole, with a median spiracle and without horny mandibles
and teeth.

Subfamily Cophylinae. The subfamily includes eight procoelous genera,
all confined to Madagascar. The prevomer, instead of being a single bone,

is divided, with the part in front of the choana separated from the part behind. The maxillary and vomerine teeth are reduced. Breeding habits are unknown, but females with large ovarian eggs have been reported in two genera, which suggests that the eggs may be laid on land.

Subfamily **Asterophryinae.** These are diplasiocoelous, or more rarely procoelous, microhylids that lack teeth and retain the large, primitive, undivided prevomer. Four genera containing several dozen species are found in the Papuan region. The eggs are laid on land, development is direct, and the embryo is much like that of *Eleutherodactylus*.

Subfamily **Sphenophryninae.** These are procoelous forms with an undivided, platelike prevomer. The teeth are nearly always absent, but some forms have vomerine odontoids. Reproduction is like that of the Asterophryinae. Five genera comprised of thirty-six or more species occur in the Papuan region and in Queensland, Borneo, and the Philippines.

Subfamily **Microhylinae.** In these diplasiocoelous, or more rarely procoelous, forms, the prevomer is much reduced and is usually confined to

FIG. 13-9. The Mexican Sheep Frog, *Hypopachus cuneus*. The common name refers to the call of the breeding male. In many of this group the call closely resembles that of a forlorn lamb.

the anteromesial border of the choana. Teeth are absent. This is the largest subfamily in the Microhylidae. It ranges from Ceylon and southern and eastern India south to the Celebes, north to Manchuria, and from the central United States to Argentina. Ten of the sixteen genera occur in the Asiatic portion of the range. There seems to be an overlapping of characteristics between the Asiatic and American forms. Reproduction is aquatic.

Subfamily Brevicipitinae. These are diplasiocoelous genera with large, platelike prevomers. Teeth are absent. Unlike members of the other related subfamilies all the genera retain a complete shoulder girdle. They are confined to southern and eastern Africa and seem to form a closely allied, compact group of four genera.

Breviceps, and probably the other members of the subfamily, lay their eggs in burrows on land. The embryo differs from that of the Asterophryinae in having an operculum and a muscular rather than vascular tail. It hatches at an advanced stage and completes metamorphosis in the nesting site.

Subfamily Melanobatrachinae. These procoelous forms have small prevomers and lack teeth. There are only three genera, confined to the mountainous regions of southwestern India and Tanganyika in Africa. *Hoplophryne* lays its eggs in the internodes of bamboos or between the leaves of wild bananas. The tadpoles are specialized for egg-eating, and superficially resemble those of the bromeliad-breeding hylas included in the suborder described next.

Family Phrynomeridae

This family of frogs apparently stands in the same relation to the microhylids as the Rhacophoridae do to the ranids. That is, they seem basically to be microhylids that have taken up an arboreal existence and have developed intercalary cartilages. They have diplasiocoelous vertebral columns, dilated sacral diapophyses, a small prevomer restricted to the anteromesial border of the choana, and a reduced pectoral girdle. The family is entirely African in distribution and contains only one genus, *Phrynomerus,* comprised of a half-dozen species widely scattered over Africa south of the Sahara. So far as the life history is known, there is a free-swimming tadpole which resembles that of the typical microhylids.

SUBORDER PROCOELA

A bewildering variety of forms, many of them highly specialized, are included in this suborder. The vertebral column is typically procoelous, though occasionally free intervertebral discs are formed. The pectoral girdle is usually arciferal, with posteriorly projecting epicoracoid horns always present. Occasionally the epicoracoids are more or less fused. The urostyle articulates with the sacral vertebra by a double condyle (sometimes urostyle and sacrum are fused). Ribs are never present. The suborder is nearly cosmopolitan in distribution and includes six recent families.

Family Pseudidae

This family includes two small genera of highly aquatic South American frogs. An extra phalanx is present in each of the digits. This increase in phalangeal formula is apparently an adaptation for swimming, similar to that of aquatic mammals. The extra element is a different structure, and serves a different function from the one found in the foot of the various groups of tree frogs. The sacral diapophyses are cylindrical, maxillary teeth are present, and the thumbs are opposable. Reproduction is aquatic; the eggs are laid in a frothy mass. *Pseudis paradoxus* is remarkable for the large size of the tadpoles. They may be more than 25 cm. long, whereas the adults are only about 65 mm. in length.

Family Bufonidae

These are the true toads, including the familiar, squat-bodied, short-legged bufos that most of us think of when we hear the word "toad." The girdle is arciferal or the epicoracoids are partially fused. The sacral diapophyses are dilated. There is a tendency toward a reduction in the number of vertebrae, apparently resulting from a forward shift of the sacral articulation, with the original sacral vertebra perhaps becoming incorporated into the urostyle. The omosternum is absent (present but cartilaginous in *Nectophrynoides*). Maxillary teeth are lacking. Bidder's organ is present in the male.

The genus *Bufo* occurs on all the major land masses of the world except Greenland, Australia, New Guinea, and New Zealand. Two genera, *Dendrophryniscus* and *Oreophrynella*, are found in South America. Twelve other genera of bufonids occur in the tropics and south temperate zone of Africa, southern Asia, and the East Indies.

Bufos typically breed in open water. Amplexus is axillary and the small-yolked eggs are characteristically laid in long strings on the bottom. As with other frogs that lay in open water, the eggs are numerous. The tadpole is short and plump-bodied—the typical polliwog. The tadpole of the Philippine toad, *Ansonia*, which breeds in mountain torrents, is depressed,

FIG. 13-10. "Which like the toad, ugly and venomous . . ." The American
Toad, *Bufo a. americanus.*

with a strong, muscular tail, reduced fins, and a suckerlike oral disc. The
African *Nectophryne* lays its eggs on land and lacks a free-swimming tadpole stage. Several of the genera (e.g., *Pelophryne* of the Philippines and
Laurentophryne of Africa) produce relatively few, unpigmented, large-
yolked eggs. Little is known of the life histories of these forms; it is possible
that they may also lay their eggs away from water. The Brazilian *Dendrophryniscus brevipollicatus,* a rough-skinned little toad of the forest floor,
is reported to breed on bromeliads.

The African live-bearing toad, *Nectophrynoides,* departs most remarkably
from the normal anuran reproductive pattern. Breeding takes place on
land and fertilization is internal. The male lacks a copulatory organ but the

opening of his cloaca is more ventral in position than that of the female. He grasps her in the axillary region and brings his cloacal opening into apposition with hers, thereby transmitting the sperm directly to her cloaca. The young go through their larval development in the ·oviduct of the mother. The number of eggs is much smaller than in the water-breeding bufos, but even so more than 100 young have been taken from a single female *N. vivipara*. Although the young are born as fully formed frogs, the embryos still retain some typical tadpole characteristics. However, they lack gills, adhesive organs, labial teeth, and horny beaks. They seem to represent a transitional stage in the evolution toward direct development.

Family Atelopodidae

Members of this family are small to medium-sized frogs, usually brightly and conspicuously marked with red, yellow, and black. The girdle is firmi-sternal, the omosternum lacking. The sacral diapophyses are dilated. Only

FIG. 13-11. *Atelopus varius zeteki,* the Golden Frog of El Valle de Anton, Panama. El Valle is now a tourist resort and this frog is one of the popular attractions there. Lapel pins modeled after it may be purchased in Panama.

two genera are included. The Variegated Toads of the genus *Atelopus* include about 25 species, which are widespread in Central and South America. *Brachycephalus* is known only from eastern Brazil. It differs from *Atelopus* in the absence of Bidder's organ and in the presence of a broad, dorsal, bony shield, confluent with the processes of the second to seventh vertebrae.

So far as is known, all atelopodids lay their eggs in water and pass through an aquatic tadpole stage.

Family Hylidae

This vigorous family of tree frogs consists of a group of procoels that have developed intercalary cartilages. They lack Bidder's organ; the sacral diapophyses range from cylindrical to strongly dilated; the pectoral girdle is arciferal; maxillary teeth are present. The family includes 32 genera and several hundred species. With the exception of the genera *Hyla* and *Nyctimystes,* the family is found only in the New World and seems to be centered in the tropics. *Hyla* has spread around the world, but it is missing from most of Africa, from all but the northernmost region of Arabia, from India, and from most of the southern coast of Asia. It is abundant in Australia and New Guinea. *Nyctimystes* occurs only in New Guinea.

Most hylids are arboreal and have enlarged digital discs for climbing, but a few, such as the Cricket Frogs (*Acris*) of the United States, have become secondarily terrestrial and possess reduced discs. Some hylids have the skin of the head fused to the skull and have developed grotesque, bony casques. Many are spectacularly colored. *Hyla heilprini,* for example, has been described as "pea-green . . . brightly variegated with gold and sky blue."

Various attempts have been made to divide the Hylidae into subfamilies, but none seems satisfactory. We simply do not know enough yet about the structure of most of the members of this family. Some hundreds of species have been described in the genus *Hyla* alone, and the result is an unwieldy and probably unnatural assemblage.

The Hylidae show a wide variety of life history patterns. Most of them, including all those found in the United States, simply lay large numbers of eggs in open water. Some species breed in still pond or lake water, others breed in streams, and the tadpoles are modified accordingly.

Other hylids have more specialized habits. The Blacksmith Frogs, *H. rosenbergi* and *H. faber,* build basins of mud in or at the edges of pools for the reception of the eggs. The tadpoles have enormous gills with which

FIG. 13-12. The Mexican Treefrog, *Smilisca baudini.*

they cling to the surface film of the small amount of water contained in
these mud nests. They metamorphose before leaving the nest.

All of the species of *Hyla* on Jamaica lay their eggs in the little water
held at the base of the leaves of bromeliads. The tadpoles are specialized
for feeding on frog eggs or other tadpoles.

Females of the Leaf Frogs, *Agalychnis* and *Phyllomedusa*, select leaves
over water on which to deposit their eggs. The spawning pair moves slowly
forward from the tip to the stalk of the leaf, folding it into a nest which
is open at both ends. When the tadpoles hatch, they fall through the
opening into the water below.

Several South American genera are sometimes placed in a separate subfamily, Hemiphractinae. Structurally, the egg-carrying tree frogs resemble
the other hylids, but they differ in that the eggs are carried in a mass on

the back of the female. The eggs of *Cryptobatrachus* and *Hemiphractus* are exposed, but those of *Gastrotheca* and *Amphignathodon* are enclosed in a pouch formed by a fold of skin. This pouch is permanent, though reduced during the nonbreeding season, in contrast to the temporary, individual pits in which the eggs of the Surinam Toad (*Pipa*) develop. The young may either leave the pouch as tadpoles and finish their development in water, or metamorphose within the pouch and emerge as young frogs.

Family Leptodactylidae

The Southern Frogs have an omosternum and, usually, a free urostyle. The girdle is typically arciferal, but one subfamily (Rhinodermatinae) has epicoracoids that are partly fused. Bidder's organ is lacking; maxillary teeth may or may not be present. The family is widely distributed in the Australian region and in Central and South America, with a few species reaching the southern United States. One genus, *Heleophryne*, occurs in Africa.

Subfamily Leptodactylinae. Members of this subfamily have an arciferal girdle. The sternum is narrow and may be divided at the posterior end.

FIG. 13-13. The Brazilian Horned Toad, *Ceratophrys dorsata.* These frogs use their huge mouths for catching other frogs on which they feed.

The sacral diapophyses are usually cylindrical. The group includes many South American frogs; a few species reach the extreme southern United States. It includes the so-called "South American Bullfrog" (*Leptodactylus pentadactylus*) and, in the West Indies, the Mountain Chicken (*Leptodactylus fallax*), which is large enough to be an important staple in the diet of natives.

The two most abundant genera, *Leptodactylus* and *Eleutherodactylus*, differ in reproductive habits. All species of *Leptodactylus* deposit their eggs in frothy nests, usually constructed in or near bodies of water. The larvae have very slim bodies and, on hatching, wriggle through the foam to reach the water. A few species have become more terrestrial. *L. marmoratus* scoops out a small basin in the ground some distance from water. The eggs and frothy mass are deposited in this basin, which is then roofed over with mud. The young hatch and pass through the tadpole stage in the nest, escaping after metamorphosis through a tiny hole left in the roof. The eggs of *Eleutherodactylus* are laid on land without protective foam. The developing embryo lacks many typical tadpole characteristics and has a large, flattened, vascular, respiratory tail. The young hatch as fully formed, tiny frogs.

Subfamily Rhinodermatinae. This is a small group of short-headed little frogs of the neotropics. The epicoracoid cartilages are partly fused and the sternum and omosternum are cartilaginous. The sacral diapophyses are dilated. There is a tendency toward fusion of the vertebrae. The sacral vertebra is particularly apt to fuse to the urostyle. When this happens, it may be hard to determine that the urostyle has the double condyle characteristic of the suborder, since the fused structure of which it is the most conspicuous part terminates in the single condyle of the sacral vertebra. The five genera included in the subfamily have an oddly disjunct distribution. *Sminthillus* is found in Cuba, *Geobatrachus* in Colombia, *Euparkerella* in Peru, *Rhinoderma* in Chile, and *Noblella* in eastern Brazil.

Sminthillus limbatus of Cuba, less than 12 mm. long, is perhaps the smallest frog in the world. During the breeding season, the female lays only one large-yolked egg on land. This hatches into a fully formed frog.

The well-known Darwin's Frog, *Rhinoderma*, has one of the most unusual breeding habits reported for any frog. The female lays 20 to 30 large eggs on land. Several males gather round and watch the clutch for 10 to 20 days, until the embryos can be seen moving inside the eggs. Then, over a period of several days, each male picks up a number of eggs with his tongue and slides them down into his vocal pouch. Here they hatch and the young pass the larval stage, emerging only after metamorphosis is complete. Although it lacks a free-living larval period, the developing frog is, for a time, completely tadpole-like, with larval body and tail proportions, closed operculum, spiracle, coiled intestine, and heavy pigmentation. Even the mouth parts are typically larval, although the beaks and labial teeth fail to harden and do not become pigmented.

Subfamily Elosiinae. This is another small group of leptodactylids, which includes only three genera—*Elosia, Megaelosia,* and *Crossodactylus.* They are found in eastern Brazil. The terminal phalanges are T-shaped; the omosternum and sternum are cartilaginous; paired, scute-like structures are present on the upper surface of each digit. One genus, *Megaelosia,* has small, bony processes on the mandibular bone of the lower jaw that form pseudoteeth. They are not true teeth, which, in frogs, are found only on the lower jaw of *Amphignathodon,* a member of the family Hylidae.

Little is known of the breeding habits of these frogs. They live in the vicinity of mountain streams, and the tadpoles may be found hiding on the bottom, under overhanging margins of pools, or behind large fragments of rock.

Subfamily Heleophryninae. The few species of leptodactylids known from Africa belong to a single genus, *Heleophryne.* The terminal phalanges are T-shaped, maxillary teeth are present, the omosternum is lacking, and the sacral diapophyses are moderately dilated. The tadpoles are found in mountain torrents and are modified for that type of habitat, as are those of the preceding subfamily.

Subfamilies Cycloraninae and Myobatrachinae. The many leptodactylids of the Australian region are placed in two separate, though closely related, subfamilies. They have dilated sacral diapophyses and a broad, usually cartilaginous, sternum. They have not been clearly differentiated from the South American forms, and there is need for a thorough anatomical study comparing them to the other leptodactylids. They have been given subfamilial rank, apparently largely because they are so widely separated geographically from other members of the family, but this is not a valid criterion on which to base classification.

The two subfamilies differ mainly in the structure of the mouth parts. The Cycloraninae have large tongues, and the prevomers and vomerine teeth are well developed; the Myobatrachinae have small tongues, and the prevomers and vomerine teeth are reduced or absent.

Life history data are lacking for many of the species. Some of the Cycloraninae are known to produce foam nests similar to those of *Leptodactylus.* The Marsh Frog, *Heleioporus eyrei,* lays its eggs in a frothy mass underground. Development proceeds within the egg until the external gills have, been lost and the operculum has developed. Hatching takes place when the nest is flooded. The Marbled Frog, *Limnodynastes tasmaniensis*, lays small eggs in a foam nest floating on any available water supply. These eggs hatch in about 48 hours. *Kyarranus* lays its eggs in sphagnum or moist earth and apparently lacks a free-swimming tadpole.

Family Centrolenidae

This small family of arboreal frogs is apparently closely related to the Leptodactylidae. As for other families of tree frogs, intercalary cartilages are present. The terminal phalanges are T-shaped, the sacral diapophyses are dilated, and the astragalus and calcaneum are fused into a single, slender bone. The pectoral girdle is arciferal, and the omosternum is lacking. Most centrolenids are small, bright green frogs. The male of *Centrolene* and some species of *Centrolenella* has a long, sharp, bony spine growing out of the proximal end of the humerus. It pierces the skin of the upper arm and fits into a pocket in the forearm. *Teratohyla* has a sharp spine projecting from the pollex rudiment. The function of these spines is completely unknown. The family includes three or four genera and ranges from Mexico to Paraguay. Breeding habits are not well known. Eggs are deposited in disclike masses on the undersides of green leaves above running water, and are apparently guarded by the male. The tadpoles fall into the water after hatching.

Collateral Reading and General Reference

Boulenger, G. A. *Catalogue of the Batrachia Salientia s. Ecaudata in the Collections of the British Museum*. 2d ed. London: British Museum, 1882. (Modern herpetology really begins with the British Museum Catalogues.)

Cochran, D. M. *Living Amphibians of the World*. New York: Hanover House, 1961.

Cope, E. D. "Batrachia of North America." *Bulletin of the United States National Museum*, no. 34, 1889. (The cornerstone of modern amphibian classification.)

Griffiths, I. "The phylogenetic status of the Sooglossinae." *Annals and Magazine of Natural History*, vol. 2, 13th series, no. 22, 1959. (In this and other papers Griffiths applies modern techniques of comparative anatomy to some basic problems of the classification of frogs.)

Noble, G. K. *Biology of the Amphibia*. New York: McGraw-Hill, 1931.

Reig, O. A. "Proposiciones para una nueva macrosistemática de los anuros."

Physis, vol. 21, no. 60, 1958. (The classification used in this chapter is based in part on that proposed by Reig.)

Das Tierreich. Amphibia: Anura I, Subordo Aglossa und Phaneroglòssa (1923); Anura II, Engystomatidae (1926); Anura III, Polypedatidae (1931). Sections I and II by F. Neiden, section III by E. Ahl. Berlin and Leipzig: Walter de Gruyter. (Synopses of the world fauna for the groups covered.)

TURTLES

SALAMANDERS are sometimes mistaken for lizards, and lizards for snakes, but a turtle can never be mistaken for anything else. The short, wide body is encased in a protective armor, the shell, which is composed of a dorsal carapace and a ventral plastron. The carapace is formed of dermal bones, usually fused to each other and to the underlying vertebrae and ribs, and covered with large, epidermal scales, the laminae (sing. lamina). The bones of the plastron apparently evolved from parts of the shoulder girdle and from gastralia, and are also covered with laminae. The laminae do not correspond in shape or size to the underlying dermal bones. Usually a bony bridge on either side, formed by an extension of the plastron, connects upper and lower shell (see Fig. 14-1).

All turtles lack teeth. Instead, each jaw is usually covered with a horny sheath, the beak, which has a sharp cutting edge, the tomium. The skull is anapsid. The shoulder girdle is unique in position—it lies beneath the ribs. The vent is a longitudinal slit and the penis is single except in the family Trionychidae, in which it is multilobed. The tail is short.

The confusion arising from the varied usages of "turtle," "tortoise," and "terrapin," is even worse than that existing between "frog" and "toad." Thus, although the terrestrial forms are usually called tortoises, the members of the terrestrial genus *Terrapene* are called neither terrapins nor tortoises, but Box Turtles. On the other hand, tortoiseshell is not obtained from the terrestrial tortoises but from the marine Hawksbill Turtle. All members of the order may properly be called "turtles." We shall restrict the name "tortoise" to the members of the subfamily Testudininae—the Giant Tortoises and their allies—and use "terrapin," not for any systematic group, but for such small to moderate sized, more or less aquatic, hard-shelled turtles as are commonly used for food.

245

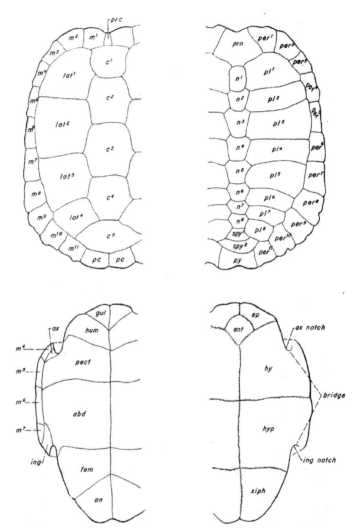

FIG. 14-1. Epidermal shell (*on left*) and bony shell (*on right*) of an emydid turtle (*Pseudemys s. scripta*). Epidermal laminae: prc, precentral; m, marginals; lat, laterals; c, centrals; pc, postcentrals; gul, gular; hum, humeral; pect, pectoral; abd, abdominal; fem, femoral; an, anal; ax, axillary; ing, inguinal. Shell bones: prn, proneural; per, peripherals; pl, pleurals; n, neurals; spy, suprapygals; py, pygal; ep, epiplastron; ent, entoplastron; hy, hyoplastron; hyp, hypoplastron; xiph, xiphyplastron; ax notch, axillary notch; ing notch, inguinal notch. Some turtles have a series of laminae known as inframarginals that separate the axillary and inguinal laminae. [From Carr, Handbook of Turtles, Comstock, 1952, by permission.]

'Many modern turtles are semiaquatic marsh dwellers and it seems probable that this has been the characteristic mode of existence throughout the course of their evolutionary history. There have been three major adaptive trends away from this mode. The tortoises have become terrestrial, other turtles are truly aquatic or even marine, and still others have adopted a bottom-dwelling existence. It is a curious anomaly that, although the primary turtle adaptation was the development of a heavy, bony box, these three adaptive trends have all involved a reduction of the shell.

Many fossil turtles are noteworthy for the thickness of the carapace bones, but such a heavy shell would be too much of a burden for a creature moving overland. Modern turtles, and particularly the larger terrestrial tortoises, have met the problem of the weight-to-volume ratio, not through a reduction of the number of bones of the shell, but through a reduction in the thickness of the individual bones. Many land turtles have also developed a high, domed shell, perhaps in part as a protection against gnawing predators that might more easily crack a flat one. The African Pancake Tortoise, *Malacochersus tornieri*, which lives in rocky terrain, has such thin bones in the flattened carapace that the shell is flexible. The animal can squeeze into narrow crevices or between boulders and inflate its body with air. It is then almost impossible to extricate.

Except perhaps for the sea snakes, the most aquatic of all living reptiles are the sea turtles, which are adapted for swift movement through the water. The Leatherback, *Dermochelys*, has undergone an extensive loss of shell bones; its carapace is simply made up of a mosaic of small dermal elements. Other marine turtles have a reduced plastron and the embryonic gaps between the lower ends of the ribs sometimes persist even to maturity. The sea turtles have also developed an efficient swimming stroke. The forefeet have become flippers and are moved with an up-and-down beat

FIG. 14-2. The African Pancake Tortoise, *Malacochersus tornieri*, which is adapted for living in rocky crevices.

similar to the motion of a bird's wing in flight. Tortoises are traditionally "Slow-and-Solid" but marine turtles are among the swifter of the tetrapods with probable top swimming speeds of at least 32 km. an hour.

Other turtles, such as the snappers, softshells, and the bizarre South American side-necked turtle, *Chelys fimbriata,* are adapted for bottom dwelling. The plastron of the snappers is greatly reduced. The softshells lack the laminae entirely. The shell is covered instead with an undivided, leathery skin whose edges conform to the bottom contour as the animal lies hidden under silt or sand.

Of all the reptiles, turtles are of the greatest direct economic benefit to man. They are probably eaten everywhere they occur. Not only do they provide the gourmet with green turtle soup and terrapin stew, but they and their eggs are a major source of protein in many parts of the world where meat is hard to come by. The dorsal laminae of the marine Hawksbill Turtle are the source of the beautifully marked tortoiseshell, used for many centuries in making jewelry and other *objets d'art.*

Turtles are conservative in their breeding habits. All species are oviparous and even the marine forms must come to shore to bury their eggs above the high tide mark.

The order Testudinata, a small group today, comprising about 335 species, is divided into two suborders—Cryptodira and Pleurodira. Cryptodira in turn contains five superfamilies. The fact that such a relatively small number of genera (about 65) can be separated into so many sharply distinct groups is indicative of the extreme antiquity of the order.

SUBORDER PLEURODIRA

The turtles of this suborder are known as side-necked turtles because they withdraw their heads by bending their necks laterally instead of vertically as do the cryptodires. The cervical vertebrae have rather high spines posteriorly and well-developed transverse processes for the insertion of the muscles that bend the neck. Their central articulations are well developed, but never double. The pelvis is fused to the plastron and sutured to the carapace. The temporal roof may be emarginated from behind and is usually more or less emarginated from below. A pair of mesoplastra (small bones lying between the anterior hyoplastra and posterior hypoplastra) are sometimes present.

These aquatic turtles are found today only in the southern continents—South America, Africa, including Madagascar, and Australia. The suborder is divided into two families.

Family Pelomedusidae

These turtles are able to tuck the head and neck into the shell so that the neck is concealed; hence they are sometimes called hidden-necked turtles. The skull is emarginated behind, nasal bones are absent, and the second cervical is biconvex. Mesoplastra are present. The family includes three Recent genera: the African *Pelomedusa* and *Pelusios* are moderate-sized forms, with maximum shell lengths of around 30 cm.; *Podocnemis* of South America and Madagascar is larger, attaining a shell length of 80 cm.

All of these turtles are aquatic, although *Pelomedusa* may wander rather freely overland. It is reported to aestivate during the African dry season. *Pelomedusa* is said to be carnivorous; *Podocnemis* is largely herbivorous.

Podocnemis expansa breeds on sandy islands midstream in the Orinoco and Amazon rivers. The females crowd together on the tiny islands, each digging a hole about 50 to 60 cm. deep and depositing 80 to 200 eggs. For years, natives of these regions have conducted highly organized egg hunts, gathering eggs by the millions and mashing them to extract the oil. The natives also prey on the hatchlings as they emerge and fish for the adults in the rivers. Only animals with a very high biotic potential could long survive such heavy predation, and the turtles are no longer as numerous as

FIG. 14-3. The Marsh Side-necked Turtle of Africa, *Pelomedusa subrufa*.

they once were. The population will probably continue to decline until it is no longer profitable to hunt for them.

Family Chelidae

The head of these turtles can be more or less withdrawn under the margin of the carapace but the neck remains exposed. The skull is little emarginated from behind, nasal bones are usually present, the fifth and eighth cervical vertebrae are biconvex, and mesoplastra are lacking. The jaw is usually long, slender, and weak. Because the neck is often longer than the carapace, these turtles are sometimes known as snake-necked turtles. Length of the carapace varies from about 15 cm. to 40 cm.

There are 10 genera of Chelidae: 4 are found in Australia and New Guinea, and 6 in South America.

One of the most bizarre creatures that ever lived is the South American Matamata (*Chelys fimbriata*). Its snout is drawn out into a snorkel-like proboscis with the nostrils at the tip, and the tiny eyes are placed far forward in the head. The mouth is very large, reaching back to the region of the ears, but the jaws are weak. The skin of the sides, lower part of the head, and long, thick neck is fringed and frayed. The carapace is rough and ridged and has three broad keels.

The female of the Long-necked Turtle of Australia (*Chelodina longicollis*) scoops out a circular hole in the ground and deposits as many as

FIG. 14-4. A South American Snake-necked Turtle of the family Chelidae, *Platemys platycephala*.

20 elongate, white eggs. They are laid in November or December and the young appear in February or March.

SUBORDER CRYPTODIRA

This suborder includes almost all the modern turtles. It is distributed on all continents though only marine forms reach the shores of Australia. The skull roof is usually much emarginated from behind. There are never any mesoplastra. The pelvis is never fused to the plastron. The cervical vertebrae are distinctive. The postzygapophyses are set wide apart. The central articulations are well developed, always broad, and typically double on the posterior cervicals. The posterior cervical spines are low and the transverse processes are almost absent. The neck is more or less retractable in a vertical position, bending in a sigmoid curve as the head is drawn back into the shell. There are five natural groups within the suborder Cryptodira.

Superfamily Testudinoidea

The three families included in this superfamily comprise the modern amphibious and terrestrial turtles. The shell is usually complete and epidermal laminae are always present. The limbs are not modified into paddles as they are in the sea turtles. There are one or two biconvex centra in the neck, and two or three of the neck joints are usually double.

Family Dermatemydidae. Of this rather primitive family, only one genus, with a single species, is living today. This is *Dermatemys mawi*, found from Veracruz, Mexico, to Guatemala and Honduras. It is a rather flat, aquatic turtle with a very short tail, and is about 45 cm. long. The proneural bone does not have elongate, rib-like (costiform) processes extending below the marginals. The plastron is well developed, not cruciform in shape, and is joined by sutures to the carapace. There is only one biconvex centrum in the neck and the eighth centrum is doubly concave in front. The tenth dorsal rib is not fused to the pleural. The alveolar surface of the maxillary bone (the crushing surface of the upper jaw) is broad and ridged.

In countries where these turtles live, natives fish for them in muddy backwaters and oxbow lakes, and sometimes sell them in the markets. Nothing has been recorded of their life history or habits.

Family Chelydridae. This small, New World family includes one of the largest freshwater turtles, the massive Alligator Snapping Turtle (*Macroclemys temmincki*) and also one of the smallest, the dwarf Stinkpot (*Sternothaerus odoratus*). The proneural bone has costiform processes which

FIG. 14-5. The "Jicotea" of Central America, *Dermatemys mawi*.

extend beneath the marginals. The tenth dorsal vertebra usually lacks ribs. The alveolar surface of the maxilla is typically broad and without ridges. There is one biconvex centrum in the neck and the eighth centrum is usually doubly concave in front. Small as is the family, it is divided into three subfamilies, each with only two living genera.

Subfamily Chelydrinae. The snapping turtles are large, with big heads, powerful jaws, long tails, small plastrons, and ugly tempers. The carapace is rough and ridged; the small plastron is cruciform and loosely joined to the carapace by a narrow bridge. An entoplastron is present; there are 11 marginal laminae on each side. The feet are webbed, and the toes long. The Alligator Snapping Turtle of the southeastern United States may reach a shell length of more than 600 mm. and a weight of 440 kg. The smaller Common Snapping Turtle (*Chelydra*), which is found from southern Canada to northern South America, seldom weighs more than 110 kg. Both forms are aquatic, though Common Snappers sometimes wander overland, and they eat both plant and animal food.

The breeding habits of the Common Snapper (*Chelydra serpentina*) are highly variable. Mating may occur from late April to November. During copulation, the male holds his position atop the female by clinging to the under edge of her shell with the claws of all four feet. He then twists his tail upward and manipulates it until contact between the vents is established.

FIG. 14-6. The Alligator Snapping Turtle, *Macroclemys temmincki,* of the southeastern United States. Note the bait on the floor of the open mouth.

The eggs are most commonly laid in June, but nesting has been reported from May to October. The nest is dug at varying distances from the water, its site apparently determined not only by the nature of the soil but also in part simply by the whim of the female. Nests have been found at distances of 1 to 25 m. from water, and are about 10 to 18 cm. deep. They apparently vary widely in form and in manner of excavation. Sometimes, at least, the cavity is definitely flask-shaped, and the excavation slants down. The female digs by alternately working her hind feet. She guides the eggs to the bottom with one of her feet and covers the nest before leaving. The number of eggs deposited varies tremendously; as few as 8 and as many as 77 (in a single nest in Manitoba) have been recorded. The eggs may vary from about 25 to 33 mm. in diameter, and while often perfectly spherical, are sometimes slightly elongated. They hatch in about three months.

Subfamily Staurotypinae. The two genera in this subfamily, *Staurotypus* and *Claudius,* are rather small turtles, with flattened shells and moderate tails. The plastron is reduced, and the bridge is narrow. The bridge of *Claudius* is connected to the carapace by a ligament; that of *Staurotypus* by a firm suture. An entoplastron is present. There are 10 marginals on each side. Both turtles are found only in Central America, and their life history is unknown.

Subfamily Kinosterninae. The little mud and musk turtles, *Kinosternon* and *Sternotherus,* have better-developed plastrons than the turtles of the other two subfamilies, but the entoplastron is lacking, a character almost unique among turtles. There are 10 marginals on each side. The anterior and posterior parts of the plastron of *Kinosternon* are nearly equal and are hinged and movable on the central part. The plastron of *Sternotherus* is smaller, and the anterior lobe is shorter than the posterior and is scarcely movable. *Sternotherus* is restricted to eastern North America north of Mexico; *Kinosternon* is found from the United States south to northern South America.

These turtles are largely carnivorous or carrion feeders, but they also take some plant food. Although highly aquatic, they sometimes travel overland and most species are fond of basking in the sun.

The Eastern Mud Turtle (*Kinosternon s. subrubrum*) typically nests in the early summer, but a late summer nesting period that may last until September has been recorded. The female searches until she finds a suitable site. She digs with her forefeet, thrusting the dirt out laterally until almost concealed, then turns around and completes the nest with her hind feet. While she is digging with the hind feet, and while laying, only her head is visible above the ground. After the two to five eggs have been deposited, the turtle crawls out and may return directly to the water, or may make some slight effort to conceal the nest cavity by levelling and scratching around the site. The completed nest is a semicircular cavity 75 to 130 mm. deep inclined at an angle of about 30 degrees. The cavity extends slightly beyond the eggs; the soil immediately around them is firmly packed, indicating that the turtle covers them carefully even though she makes no effort to conceal the entrance to the nest. Rains, pounding on the loose, sandy soil, soon obliterate all sign of it. Apparently, the eggs usually overwinter and hatch the next spring, for hatching eggs have been plowed up in April. In the laboratory, hatchlings have emerged in late September, but presumably in natural surroundings they would have overwintered in the nest.

Family Testudinidae. This is the one really flourishing group of modern chelonians, found on all continents except Australia. Most of the familiar pond and swamp turtles, as well as the terrestrial tortoises are members of this family. The carapace and plastron are unreduced and the bridge between is well developed. The proneural is without lateral costiform processes. There are usually two biconvex centra in the neck and the eighth centrum is typically doubly convex in front. The alveolar surface of the upper jaw may be broad or narrow and frequently bears one or more ridges. The family is divided into three subfamilies.

Subfamily Platysterninae. Platysternon megacephalum, the only member of this subfamily, is a rather small (carapace length about 150 mm.) turtle with a big head, long tail, strongly hooked jaws, and depressed shell. The plastron is connected to the carapace by ligaments, and the inframarginal row is complete. There are 11 marginals on each side. The temporal region of the skull is almost completely roofed. The digits have three phalanges and are not quite fully webbed.

These turtles inhabit mountain streams in southeastern Asia. Rather surprisingly, they are agile climbers and may be found in trees or on rocks, hunting for food or sunning themselves. A captive specimen ate snails and worms. The female lays only two eggs at a time.

Subfamily Emydinae. This most extensive group of living turtles includes both aquatic and terrestrial forms. Most are medium sized, but some are among the largest of the non-marine turtles. The temporal region of the skull is not roofed over. The middle digit usually has three phalanges and the toes are usually more or less webbed. The inframarginal series of laminae is never complete; there are 11 marginals on each side.

The Emydinae includes about 25 genera; representatives are found on all continents except Australia and Africa south of the Sahara. The richest and most varied fauna is that of eastern and southeastern Asia, where there are 17 or more genera. Europe has only two genera, *Emys* and *Clemmys*, and western North America a single species, *Clemmys marmorata*. The subfamily is well represented in eastern North America, where it includes such well-known forms as the terrestrial box turtles (*Terrapene*) whose plastron is hinged so that the shell can be closed completely, and the succulent Diamondback Terrapin (*Malaclemys*).

Food habits within the group are varied. Some species are strictly herbivorous, others are carnivorous, and still others seem to prefer a mixed diet.

Some of the emydines have developed elaborate courtship patterns (see p. 180). Most species lay relatively few (less than 12) rather large eggs, but some of the larger forms may deposit 30 or more at a time. *Batagur baska* is a large (shell length nearly 60 cm.) aquatic species found in estuaries, slow-flowing rivers, and canals of southeastern Asia. Around the mouth of the Irrawaddy River, egg laying takes place during January and February. The turtles come ashore every afternoon to gather in large herds and sun themselves on the sand. At night the females dig holes 45 to 60 cm. deep in the sand above the high tide mark and deposit 10 to 30 eggs about 75 mm. long. Each female lays between 50 and 60 eggs, in three batches, over a period of about 6 weeks. The incubation period has been reported to be about 70 days.

Subfamily Testudininae. These are the true land tortoises, including the lumbering giants of the Galápagos and Seychelles Islands. The feet are

FIG. 14-7. Hermann's Tortoise, *Testudo hermanni,* of southern Europe. Breeding males of this and other species of *Testudo* exercise their voices in the spring.

club-shaped, short and broad, with not more than two phalanges in any digit. The toes are completely unwebbed. The hind legs are like those of an elephant, cylindrical and columnar. The skull roof is incomplete posteriorly; the inframarginal series is never complete; there are 10 or 11 marginals on each side. Usually the shell is high and dome-shaped. Not all tortoises are large—the little *Testudo leithi* reaches a shell length of only 120 mm. The largest *Testudo* recorded, a specimen from Aldabra Island, measured 1,400 mm. in straightline shell length and weighed 254 kg.

Seven recent genera are recognized in this subfamily. By far the largest and most widely spread is the genus *Testudo,* found in southern Europe, Asia, Africa, and South America, as well as on islands of the Indian Ocean and the Galápagos Islands off South America. North America has only one testudine, the Gopher Tortoise, *Gopherus,* of southern United States and Mexico. The remaining genera are all confined to Africa and Madagascar.

In general, each tortoise has an individual resting place—a burrow it has dug, a cranny in a rock, or a sheltered nook under a plant—from which it emerges to graze during the day or, in hot weather, in the cool of the evening. Tortoises are largely herbivorous, but *Testudo graeca* has been réported to take insects, molluscs, and worms. One captive Galápagos tortoise was seen to catch two rats and a pigeon, and ate raw meat greedily. Tortoises that live in desert areas, or on rocky islets where no fresh water is available, are able to get along without drinking; other species drink copiously and enter water freely.

Courtship of tortoises seems to consist mainly of the male pursuing the female and butting her with his shell. Copulating males are surprisingly vocal—the "voice" has been described for various species as "a muffled, whistling cry," "a peculiar grunting noise," and "the intermittent winding up of a metal spring." In spite of their relatively large size, most female tortoises lay few eggs in a single clutch, usually less than seven. Some species may lay only a single egg per clutch. As with most other turtles, the eggs are buried in a hole which the female digs with her hind legs.

Superfamily Chelonioidea

These are large, sometimes enormous, marine turtles which have epidermal laminae, a heart-shaped carapace, the limbs modified as paddles, and the skull roof more or less complete posteriorly. There is only one biconvex vertebra in the neck, and the head cannot be withdrawn completely into the shell. The superfamily contains only one living family.

Family Cheloniidae. The family includes four living genera: *Caretta*, the massive Loggerheads which may weigh as much as 400 kg.; *Chelonia*, the succulent Green Turtles, so called from the color of their fat; *Eretmo-*

FIG. 14-8. The Atlantic Hawksbill, *Eretmochelys imbricata*, the source of tortoiseshell *objets d'art*.

chelys, the Hawksbills, source of the beautiful tortoiseshell; and *Lepido-chelys,* the Ridleys, the smallest of the marine turtles, with a maximum shell length of 790 mm. These turtles are tropicopolitan in the warm seas of the world; occasional stragglers are carried by warm currents as far north as Newfoundland and Scotland.

Green Turtles were once important commercially as a source of food, but heavy predation by man, on the eggs as well as on the adults, has so reduced their numbers that relatively little turtle meat is obtained for export today. Nevertheless, marine turtles remain a major source of food to natives in many parts of the world. The recent development of the plastic industry has nearly eliminated the commercial use of tortoiseshell.

Adult Green Turtles are largely herbivorous, browsing on submerged vegetation in shallow, offshore waters. Loggerheads, on the other hand, are carnivorous, and the Hawksbills and Ridleys are omnivorous.

For a species of animals to survive, its reproductive rate must be sufficient to allow enough of the offspring to reach maturity so that the breeding population is maintained. The number of eggs laid by each female gives some clue to the extent of the hazards faced by the developing young. Marine turtles lay many more eggs than freshwater and land forms. During the spring and summer months, the turtles gather offshore from the breeding beaches. At night the female leaves the sea and heaves her heavy body laboriously across the sand to above the high tide mark. Here she digs a nest and deposits her eggs. Two hundred eggs have been found in *Chelonia* and *Eretmochelys* nests, and even the smaller *Lepidochelys* may lay as many as 135 eggs. Also, the females apparently deposit more than one clutch during a breeding season—seven emergences for egg laying in two weeks have been reported for a female *Chelonia,* although probably two or three clutches is more usual for marine turtles.

After the eggs are deposited, the female fills the hole, then lumbers back to the safe haven of the sea. The males are waiting offshore, and copulation takes place, so that the eggs for the next breeding season may be fertilized.

The clutches of marine turtles are concentrated in a small area above high tide mark on the laying beaches. Hordes of predators gather to dig them up, the most destructive being man and feral dogs. After hatching, the baby turtles claw upward through the sand and head for the water. On their journey across the beach, they must run the gauntlet of hungry crabs, mammals, and hovering sea gulls. Those that survive are met in the water by swarms of predatory fish. Yet marine turtles have persisted for millions of years (*Caretta* and *Chelonia* are known from the Upper Cretaceous). The marked decline in turtle populations in recent years probably results largely from the activity of human hunters in waylaying the females on the beaches before they deposit their eggs.

Superfamily Dermochelyoidea

These marine turtles lack epidermal laminae and have the dermal bones of the carapace and plastron largely replaced by a mosaic of small platelets set in a leathery skin. The limbs are paddle-shaped and without claws, the anterior ones being very large and the posterior ones, of the adults, broadly connected to the tail by a web. The skull roof is complete; there is only one biconvex vertebra in the neck. The superfamily includes only a single family.

Family Dermochelyidae. A single species, *Dermochelys coriacea*, the Leatherback, is the only living member of the family. This extraordinary creature, one of the most remarkable of all living reptiles, can be confused with nothing else. It is distinguished from all other sea turtles by the scaleless black skin of its back and by the seven narrow ridges, formed by enlarged platelets of the dermal mosaic, that extend down the length of the back. Five similar keels occur on the ventral surface. There is a strongly marked cusp on each side of the upper jaw. This, the largest of the turtles living today, may reach a weight of 680 kg. or more, but such giants are rare. The large specimens encountered from time to time along the coasts today probably weigh around 360 kg., about the size of a large loggerhead.

Remarkably strong and rapid swimmers, the pelagic leatherbacks are widely distributed, though usually scarce, in tropical seas, and occur sporadically in temperate waters. They feed on crustaceans, molluscs, and small fishes, as well as on marine plants.

Like the females of other marine turtles, the female leatherback must deposit her eggs on land. With her front flippers she first excavates a broad crater in which her body rests and within that a nest hole. The combined depth of crater and hole may be 90 to 120 cm. After egg deposition, the female fills in hole and crater, then plows up a broad area of the surrounding sand thus effectively concealing the eggs. Between 80 and 130 eggs are deposited at one time, and there may be two or three clutches during a breeding season.

Superfamily Carettochelyoidea

The epidermal laminae of these turtles are reduced or absent, but the underlying bony shell is complete. The limbs are modified as paddles, the

FIG. 14-9. The Hinged Softshell, *Lissemys punctata*, of India.

digits are long and connected by webbing, and there are only two claws on each foot. The skull roof is emarginate behind, rather than roofed over as in the preceding group. There is one biconvex centrum in the neck. Here again this superfamily is represented by only one family.

Family **Carettochelyidae.** The sole surviving member of the family is *Carettochelys insculpta*, the Pitted Shelled Turtle. Laminae are absent and the pitted shell (about 50 cm. long) is covered only by a thin layer of soft skin. This rare and little known freshwater turtle has been found only in the Fly River of New Guinea. Its life history is completely unknown.

Superfamily Trionychoidea

The softshelled turtles lack epidermal laminae and their bony shell is somewhat reduced. The plastral bridge is absent. The shell is low, usually nearly circular in outline, and covered with a leathery skin. The neck is long and retractile, the lips are fleshy, not covered with horny beaks like those of other turtles, and the snout is drawn out into a fleshy proboscis with the nostrils at the tip. The limbs are paddlelike, with three claws on each foot. The skull is deeply emarginate behind and there are no biconvex centra in the neck. Only one family is known.

Family Trionychidae. These extremely aquatic, bottom-dwelling turtles are found chiefly in rivers, less often in ponds and swamps. They are moderate to rather large in size, with shell lengths of from 23 to 80 cm. Most species are carnivorous, waiting for their prey concealed in the bottom mud or sand, or actively foraging. Occasionally plant food is taken. Soft-shelled turtles are widely esteemed as food. They are probably also useful as scavengers. Like other turtles with reduced shells, many, though not all, of the species are short tempered and quick to bite. The living forms are divided into two subfamilies.

Subfamily Trionychinae. The typical Softshells have the hyoplastron distinct from the hypoplastron and are without cutaneous femoral flaps on the plastron. There are four genera. The widespread *Trionyx* is found in North America, Africa, and Asia. *Dogania, Chitra,* and *Pelochelys* occur in southeastern Asia, with the last extending to New Guinea.

The breeding habits of the Eastern Spiny Softshell, *Trionyx s. spinifer* of North America are probably typical for the family. Nesting occurs in June and July. The female digs a flask-shaped hole, 100 to 250 mm. deep

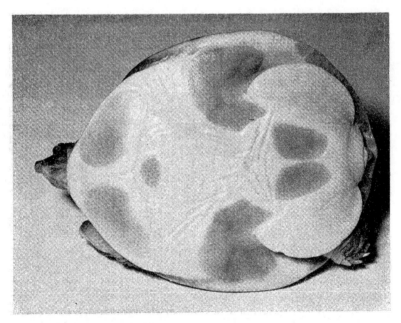

FIG. 14-10. Ventral view of the Hinged Softshell, showing pelvic flaps characteristic of this subfamily.

and 75 to 125 mm. in diameter, with a narrow neck. It takes her about 40 minutes to dig the nest. She deposits a few eggs at a time, arranges them with her feet, and rakes some earth down in the hole with them, gently packing it in, then lays more eggs. This is continued until the nest is finally completed. She apparently makes no effort to conceal the nest after she has covered the last of the eggs. The 10 to 25 eggs are about 25 to 27 mm. in diameter but are not perfectly spherical. The shell is thick and not particularly brittle. The incubation period in this subspecies is not known but eggs removed from an adult female of the Florida race (*T. f. ferox*) hatched in 64 days.

Subfamily Lissemyinae. Members of this subfamily have the hyoplastron and hypoplastron fused. A pair of strong, hinged, cutaneous flaps at the rear of the plastron close over the hind limbs when they are withdrawn. Of the three genera in the subfamily, *Lissemys* is found in India and Burma, and *Cyclanorbis* and *Cycloderma* in Africa. Their breeding habits probably do not differ greatly from those of members of the preceding subfamily.

Collateral Reading and General Reference

Agassiz, L. *Contributions to the Natural History of the United States.* Vols. 1–4. Boston: Little, Brown and Co., 1857. (A classic early American work.)

Boulenger, G. A. *Catalogue of the chelonians, rhynchocephalians, and crocodiles in the British Museum.* London: British Museum, 1889.

Carr, A. F. *Handbook of Turtles.* Ithaca: Comstock, 1952. (Most readable, and really more than a handbook. Part I is an excellent introduction to the general biology of turtles; Part II is a systematic account of the turtles of the United States and Canada.)

Pope, C. *Turtles of the United States and Canada.* New York: Alfred A. Knopf, 1939. (A sound, well-written introduction to the study of turtles, particularly those of the United States.)

Smith, M. A. *Fauna of British India. Reptilia and Amphibia.* Vol. I. *Loricata, Testudines.* London: Taylor and Francis, 1931. (An exceptionally sound faunal report.)

Wermuch, H. and R. Mertens. *Schildkröten, Krokodile, Brückenechsen.* Gustav Fischer Verlag, Jena, 1961. (A recent checklist, worldwide in scope.)

CHAPTER

LIZARDS

"ORDER SQUAMATA: Epidermal scales present; vent a transverse slit; male copulatory organs paired hemipenes; no gastralia; teeth either pleurodont or acrodont; vertebrae usually procoelous."

This terse, dry description gives no hint of the extraordinary interest the Squamata have held for man, reputedly since Eve first met the snake in the Garden. The order includes about 5,700 different species of the animals commonly known as lizards and snakes, two rather closely related groups that are classified as suborders of the large order Squamata. One suborder, Lacertilia, includes the lizards. In defining it, we must once again depend on a combination of characters, some of them lacking in a few lizards, and some of them also found in a few snakes.

Lizards have the two halves of the lower jaw firmly united at a mandibular symphysis and the size of the mouth opening is restricted; the tongue is well developed. Most lizards have two pairs of limbs, but some have lost one or both pairs, although traces of the girdles remain in all except some Typhlopidae. Usually lizards have visible external ear openings and movable eyelids with a nictitating membrane (third eyelid). A urinary bladder is typically present. The tail of many lizards is readily detachable and can be regenerated.

As a group the lizards have prospered and now number nearly 3,000 different species. Although most numerous in the tropic regions of the world, they have successfully invaded all of the continents (except Antarctica) and in Europe reach up to the Arctic Circle. They have undergone extensive adaptive radiation and have taken on a bewildering variety of habits. Some spend much of their time in water, either fresh or salt, while others have become well adapted to life in arid deserts.

The desert-dwelling forms are for the most part flat with depressed bodies which makes it easy for them to hide beneath stones and creep into narrow fissures. Many rock-haunting species have developed special pads on the toes which enable them to run swiftly over smooth surfaces. With the advent of man, some of these have shifted their quarters to the walls of houses, where an abundance of insects assures a steady food supply. Lizards for the most part lack webs between their toes; the lizard with the most strongly webbed feet (*Palmatogecko rangei*) is found in the arid deserts of southwest Africa where the webbing assists it in maneuvering in the shifting sands.

Some of the ground-inhabiting lizards are large and sluggish. The giant Dragon of Komodo (*Varanus komodoensis*) probably reaches 3 m. in length. Others are small and agile; some of them run rapidly on two legs, much as small dinosaurs did in the Mesozoic. Still other ground lizards have lost their limbs, and glide through the grass like snakes.

Some lizards have become burrowers. Their limbs are not used in digging through the soil, but are held close to the sides and are usually greatly reduced. A few have lost the limbs entirely.

Arboreal lizards have compressed bodies which make them inconspicuous as they bask on the branches of trees. Most spectacular are the many species of Flying Dragons (*Draco*) of southeast Asia and the East Indies. A Dragon may be recognized immediately by the winglike expansions of skin on either side of the flank, supported by half a dozen of the hindermost ribs. These ribs function in much the same way as the ribs of a parasol, for when the lizard is at rest it presses them against its body so that the skin is folded up and is scarcely noticeable. These expansions are simply gliding surfaces and cannot be used for true flight. The lizard rests on a tree in a vertical position and when it sees an insect flying by, it leaps into the air, soars down, catches the insect, and lands with an action so rapid that the opening and closing of the "wings" is scarcely perceptible. One has been seen to glide smoothly and steadily for a distance of 18 m.

The lizards are grouped into six infraorders.

INFRAORDER GEKKOTA

The geckos and their allies are mostly either stoutly built, short-tailed, little lizards, largely nocturnal and well adapted for climbing, or else snakelike forms with very reduced limbs. Most geckos are small creatures less than 150 mm. long. They lack an upper temporal arch, a postorbital arch, and lachrymal, squamosal, and postorbital bones in the skull. The jugal bone is small and sometimes absent. The frontal bones are usually united and surround the forebrain. Pleurodont teeth are usually present on the marginal

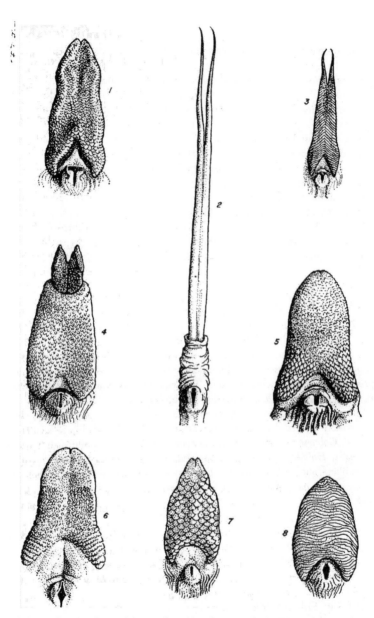

FIG. 15-1. Tongues of Lacertilia. 1. *Mabuya carinata* (Scincidae). 2. *Varanus monitor* (Varanidae). 3. *Tachydromus sexlineatus* (Lacertidae). 4. *Ophisaurus harti* (Anguinidae). 5. *Calotes versicolor* (Agamidae). 6. *Gekko gecko* (Gekkonidae). 7. *Nessia monodactyla* (Scincidae). 8. *Dibamus novae-guineae* (Dibamidae). [After Malcolm Smith.]

bones (premaxillaries, maxillaries, and dentaries) but palatal teeth are absent. The eyes have either eyelids or "spectacles" which are simply eyelids that have fused and developed transparent areas through which the lizard sees. (In one family, the eyes are reduced and covered by the skin.) The tongue is thick, fleshy and only very slightly cleft, if it is cleft at all. Most species have a postanal sac opening to the outside on each side of the base of the tail just behind the vent. In the male, a bone lies free just below the skin in front of the opening of the sac on each side. The function of these sacs is unknown. There are usually four or more transverse rows of belly scales per body segment. This infraorder includes those lizards that have well-developed voices.

The Gekkota includes five families. Three of them comprise those lizards with well-developed limbs commonly known as geckos throughout the world. In the other two, the scale-foots, the hind legs are represented by scarcely noticeable flaps on either side of the vent and the front legs are lacking.

Family Eublepharidae

The Ground Geckos differ from all other geckos in having true eyelids that can be opened and closed rather than spectacles. The digits are straight, not angularly bent at any of the joints, and are without friction pads for climbing. Postanal sacs and bones are present and the male usually has preanal or femoral pores or both. The vertebrae are procoelous. The tail is short, usually swollen, somewhat carrot-shaped.

The geographic distribution of members of this family is peculiar. They are widely scattered in desert regions throughout the world. The Banded Gecko, *Coleonyx*, is found from the southwestern United States to Central America; *Holodactylus* is in Somaliland in East Africa; *Hemitheconyx* in Somaliland and in West Africa; *Eublepharis* in Southwest Asia and on islands off the east coast of Asia; and *Aeluroscalabates* in Malaya, Sumatra, and Borneo.

Like the true geckos, and unlike most other lizards, the ground geckos are nocturnal, hiding by day under rocks or in burrows in the sand, coming out at night to forage for insects. Some, at least, have loud voices. Little is known of the breeding habits, but apparently, like most other geckos, the female lays two eggs at a time.

Family Sphaerodactylidae

This is a compact little family of geckos with procoelous vertebrae. The eye is covered by a spectacle. Simple pads are present on the toes of one genus (*Sphaerodactylus*); all of the others lack pads. Toes may be straight or angularly bent. Postanal sacs and bones as well as preanal and femoral pores are lacking. This family includes the tiniest of the lizards, some of the *Sphaerodactylus* being less than 50 mm. long.

In contrast to the Eublepharidae, the Sphaerodactylidae are restricted in distribution to the tropics of the New World. They tend to be more active during the day than most geckos, and more terrestrial. Also in contrast to other geckos, they are apparently voiceless, and the female, instead of laying two eggs, lays but one hard-shelled egg at a time.

Family Gekkonidae

The true geckos form a large tropicopolitan family of about 60 genera and some hundreds of species. Most are stoutly built little lizards, nocturnal and arboreal, with big, spectacled, cat-like eyes, the pupil contracting to a slit in daylight and opening wide at night. The vertebrae are amphicoelous. The digits frequently have both claws and friction pads for climbing. The males usually have preanal and femoral pores, as well as postanal sacs and bones.

Many geckos have attached themselves to the dwellings of men. This really is a type of mutualism, since the insects that are also attracted to houses provide food for the geckos and the geckos, by feeding on the insects, help keep these pests under control. Man has been slow to recognize his debt. Indeed, superstitions about them are as widespread as the geckos. It is claimed that these harmless little animals are highly venomous, that their bite is fatal, that they poison man's food and drink, that they can cause leprosy by running over the face of a sleeper. As with the rat and mouse, man has unwittingly transported geckos about the globe so that the original distribution of many of the genera may never be known.

Few lizards can compete with the geckos in their ability to discard the tail. This defense mechanism is so marked in some species that it is nearly impossible to find an adult specimen with the original tail intact.

The name gecko probably arose as an attempt to imitate the call of some species of these lizards. Their ability to vocalize is remarkable, for most lizards are silent creatures. The sound is perhaps produced by clicking the broad tongue against the roof of the mouth. It has been variously transcribed as "checko," "tocktoo," "toki," "tok," or "chick chick."

Perhaps most spectacular to people seeing geckos for the first time is their ability to run over a window pane or up a smooth vertical wall and across the ceiling. The geckos capable of this sort of climbing have some part of their digits dilated to form adhesive discs. In the most arboreal

FIG. 15-2. The underside of the Moorish Gecko, *Tarentola mauretanica*, seen as it clings to a sheet of glass.

forms the underside of the disc is made up of a transverse or fan-shaped series of narrow plates bearing minute, hairlike processes or papillae, which can be pressed into tiny irregularities of the surface. The pad does not function well on a moist surface. Indeed, a convenient way to collect geckos is with a water pistol. If one squirts the wall on which a gecko is climbing, it will fall alive and uninjured to the ground.

Most geckos are gentle little creatures, but not the Tokay Gecko of the Dutch East Indies and the Philippines. When annoyed, the Tokay inflates its body and hisses and puffs loudly, holding its jaws wide open in readiness to attack. If the provocation continues, the lizard rushes forward and seizes some part of its annoyer in its powerful jaws, hanging on with bulldog tenacity. The Tokay is the largest of the geckos, with a head and body length of about 175 mm. and a total length of over 300 mm.

Except for two New Zealand genera which bear living young, all geckos lay two eggs with shells that are hard and brittle instead of being leathery like those of most other lizards. Sometimes more than two are found in a single spot, but this is because two or more females have chosen the same site for egg deposition.

Family Pygopodidae

At first glance, the snake-lizards bear little resemblance to the geckos, but they are very similar to them in many structural characters so that it seems best to include them in the infraorder Gekkota. They are slender and snakelike, with no front legs and for hind legs only a pair of small, scaly flaps near the vent. Vestiges of the pectoral girdle remain, and from one to four digits can be recognized in the hind limb. The vertebrae are procoelous, postanal sacs and bones are present, the eye is covered with a spectacle, and the pupil is vertical. The animals are between 150 and 750 mm. long. The tail is considerably longer than the head and body and is very easily shed.

There are about 8 genera and 14 species of pygopodids, some of them known from only one or two specimens. They are found only in the Australian region. All are nocturnal, some are also fossorial, and little has been recorded of their habits. Most are apparently insectivorous but some of the larger species feed on other lizards. So far as is known, all are oviparous.

Family Dibamidae

The family comprises a single genus of slender and worm-like lizards not more than 225 mm. long. As in the pygopodids, only vestiges of the pectoral girdle remain. The hind limbs too have disappeared in the female but they are represented by scaly flaps in the male. The Blind-skink (*Dibamus*) differs from the pygopodids in having the eye greatly reduced and covered by skin. The tail is short and apparently cannot be shed. The egg has a calcareous shell. *Dibamus* ranges from southern Indochina and the Philippines to the New Guinea archipelago.

INFRAORDER IGUANIA

This infraorder, and the Scincomorpha, include the majority of the typical lizards, the animals most people think of when they hear the word. The iguanians are many and varied, some large, others quite small, frequently brightly colored, and often ornamented with crests, spines, frills, and throat fans. They are diurnal and may be either arboreal or terrestrial. *Amblyrhynchus* of the Galápagos Islands is semimarine. None shows any tendency toward the development of a snake-like body form or reduction of the limbs.

The temporal arch is present, the skull is high, the teeth are either pleurodont or acrodont, the parietals are fused to form a single bone. There are six cervical vertebrae, and four or more rows of transverse belly scales per body segment. The tongue is simple rather than divided into anterior and posterior portions. The eyelids are well developed, the pupils round. Femoral pores are occasionally present. The tail is not very fragile, although breaking point septa are present in the caudal vertebrae.

This infraorder contains two very similar families—the Iguanidae and

Agamidae—which are essentially New and Old World counterparts of each other.

Family Iguanidae

This is the largest family of lizards of the New World, to which it is almost restricted. There are two genera (*Chalarodon* and *Oplurus*) in Madagascar and one (*Brachylophus*) in Polynesia, on the Fiji and Tonga Islands, but there are more than 50 genera in the western hemisphere. These lizards have a pleurodont dentition in which the teeth are attached on the inner surface rather than along the dorsal margin of the lower jaw. In relation to this, the splenial bone of the lower jaw is well developed as compared to its reduced condition in the sister family of the Agamidae.

The iguanids range in size from tiny forms less than 125 mm. long to the giant iguanas which may be 1,800 mm. long. These large iguanas are used as food in some tropical countries. The smaller iguanids are mostly insectivorous or carnivorous, but many of the larger ones are herbivorous. The marine iguanas of the Galápagos enter salt water to forage for seaweed, which seems to be their favorite food.

A characteristic habit of the family is head bobbing which, in such forms as *Anolis,* is correlated with the distention of a throat fan. The head is raised above the surface and bobbed backwards so that the brightly colored fan is spread, but the characteristic bobbing motion occurs also in iguanids that do not have a throat fan.

Iguanids are usually oviparous, but some species of the Horned Lizards

FIG. 15-3.   The Iguana, *Iguana iguana.*

FIG. 15-4. The Key West Anole, *Anolis sagrei stejnegeri.* The bifid tail apparently resulted from an injury.

(*Phyrnosoma*) and some of the Spiny Lizards (*Sceloporus*) are ovoviviparous. The eggs are soft-shelled and are frequently buried in the ground, even by species that are largely arboreal.

Family Agamidae

This large family is the Old World counterpart of the New World Iguanidae. They are most numerous in the Oriental Region, but are found also in southeastern Europe, Africa, Australia, and the New Guinea archipelago. They are absent from the only places in the Old World from which iguanids are known, the Fijis, Tongas, and Madagascar. Thus the two families are entirely separate in geographic distribution.

The Agamidae differ from the Iguanidae only in the structure of the jaws and teeth. The teeth are acrodont, that is, they are attached to the upper margin of the jawbone, and the dentition is heterodont, with the teeth divided into incisor-like, canine-like and molar-like ones. We say "like" since it is not at all certain that these are homologous to the similar teeth in mammals. Many species have well-developed ornamental crests, frills, or throat pouches, frequently brilliantly colored. The "wings" of the flying dragons are as bright as those of butterflies. Most agamids are moderate in size. *Hydrosaurus,* the water lizard of the East Indies and New Guinea, may reach a length of 900 mm. At the other extreme, the Toad-headed Agamids, *Phrynocephalus,* may be less than 125 mm. long.

Like the iguanids, the agamids have undergone adaptive radiation; some

FIG. 15-5. This Mastigure *Uromastix acanthinurus werneri*, refused food until offered the zinnia, whereupon it bit off the petals as fast as it could swallow
them. It is a resident of northern Africa.

are terrestrial, some arboreal, some verge on fossorial. Most are carnivorous,
but some are omnivorous and one, *Uromastix*, is largely herbivorous.

The Agamidae seem to be mostly oviparous but at least two genera,
Phrynocephalus of Central Asia and *Cophotes* of Ceylon, are ovoviviparous.
The Flying Dragons, although otherwise entirely arboreal, do come down to
bury their eggs in the ground.

INFRAORDER RHIPTOGLOSSA

This infraorder contains the most weird-looking of all lizards, the true
chameleons. They are highly modified for arboreal life, with short, compressed bodies, and coiled, prehensile tails. The skull may be ornamented
with horns, crests, and tubercles, and is produced backward to form a
grotesque, helmetlike casque. Most chameleons are moderate in size, the
various species ranging from less than 150 mm. to 600 mm. in total length.

The dentition is acrodont, the dorsal temporal arch is present, the parietals
are fused to form a single bone. There are only three cervical vertebrae.
The tongue is much elongated and is enlarged at its distal end, but not
divided into distinct fore and aft sections. There are four or more rows of
transverse scales on the ventral side of the body for each body segment. The
tail lacks breaking points. The thick, granular eyelids are united except for a
small slit in the center. The hands and feet are zygodactylous (with yoked
digits). The two inner fingers of the hand are bound together in a bundle
which opposes the similarly joined three outer fingers. The three inner toes
of the feet oppose the two outer toes. This forms a very efficient grasping
mechanism.

Family Chamaeleonidae

This is the only family in the infraorder. It is divided into 6 genera; more than 80 species are recognized. The vast majority live in Africa and Mada-. gascar, but the Common Chameleon of northern Africa (*Chamaeleo chamaeleon*) is also found on some of the islands of the Mediterranean. Two species occur in southwestern Asia and one inhabits the Indian region.

Most spectacular is the ability of chameleons to move each of their round, protuberant eyes independently. This ability is slightly developed in some iguanians, but only in the true chameleons has it reached the point where one eye is entirely independent of the other. It is fascinating to watch a chameleon resting on a limb, with the left eye rolling in one direction while the right eye rolls around peering in the other. When some large insect or small bird is spied, both of the eyes come to focus on it, and the lizard creeps along with painful slowness until within a few inches of its victim. Then, faster than the human eye can follow, the long, extensile tongue shoots out, grasps the creature, and pulls it back into the mouth, and with a gulp the prey is gone.

Chameleons are famous for their ability to change color, from white through shades of yellow, green, and brown, to black, variously spotted and blotched with contrasting colors. Some iguanians, such as *Anolis*, show a similar ability—which is why anoles are often called chameleons. These color changes are responses to changes in light, heat, and the emotional state of the animal, not, as is popularly believed, to the color of the background.

Some chameleons are oviparous, others ovoviviparous. A pair of captive *Chamaeleo zeylanicus* from India mated during the first week in October. The female would not allow the male to approach her after mating had occurred. On the 9th of November, a little more than one month after copulation, she descended to the ground and began digging like a terrier, packing the loose earth with her forelegs and kicking out behind with her hind legs. She remained in the hole overnight and continued to work on the nest the following day. About 2:00 o'clock in the afternoon of the 11th, she emerged and spent the rest of the day pulling back the earth with her hands and ramming it well behind her with her hind legs. She did not finish filling the nest until the next morning. The 31 oval, soft-shelled eggs measured 13 by 7 mm. and were buried about 30 cm. below the surface of the ground. Eggs as large as 19 by 12 mm. have been found in dissected specimens.

INFRAORDER SCINCÓMORPHA

This is a large, cosmopolitan group of lizards, moderate to small in size, with a strong tendency toward reduction of the limbs and development of a snake-like body habitus. The degenerate forms are usually burrowers. In the Scincomorpha, the dentition is pleurodont, the dorsal temporal arch is usually present, and the parietals are fused. The tongue is simple. There are six cervical vertebrae and fewer than four rows of transverse belly scales per body segment.

The infraorder is cosmopolitan in distribution and comprises six families.

Family Scincidae

The abundant and ubiquitous true skinks are mostly secretive, semi-burrowers, small in size, with highly polished scales. Most are less than 200 mm. long, and the largest, *Corucia zebrata* of the Solomon Islands, is only about 600 mm. in length. The temporal openings of the skull are more or less roofed by backward growths of the postfrontals. Pterygoid teeth are often present. The limbs may be present or absent; those species that have lost their limbs always possess some traces of the pectoral and pelvic girdles. Abdominal and parasternal ribs are sometimes present, chiefly among the burrowing forms. The head, body, limbs, and tail are protected by osteo-derms. The head is covered with symmetrical shields, the pupil is round, femoral pores are absent. The tail is fragile and when broken off is quickly regenerated. Some of the skinks have developed a transparent window in the lower eyelid which enables the lizard to see while the eyelids are tightly shut. The most extreme development along this line is found in the little, active, snake-eyed skinks (*Ablepharus*) whose eyelids are wholly trans-parent and immovable, permanently covering the eyes like a pair of watch glasses. Not all skinks are smooth and shiny; a few bizarre-looking forms have big, spiny scales on their backs, sides, and tails. The stout-bodied, stubby-tailed Shingle-back of Australia (*Trachysaurus*) looks like an ani-mated pine cone.

Although they are cosmopolitan in distribution, skinks are most numerous in Australia, the islands of the western Pacific, the Oriental Region, and Africa, and are poorly represented in the Americas. There are nearly 50 genera, and more than 600 recognizable species.

The vast majority of skinks are terrestrial in their habits and are usually extremely active; a few show arboreal tendencies, but none is so highly

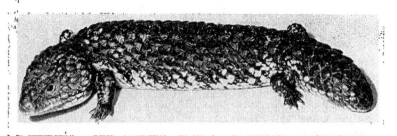

FIG. 15-6. The bizarre Shingle-back of Australia, *Trachysaurus rugosus*. The short tail is responsible for the common name "stump tail" which is often applied to these skinks.

adapted for climbing as are members of the preceding infraorders. Some skinks have become rock dwellers and a few are good swimmers. They are often found under drifts of dead leaves, piles of coconut husks, and rotting vegetation or decaying logs. The terrestrial skinks are diurnal whereas the burrowers seem to be largely crepuscular or nocturnal. Most skinks are insectivorous, but some of the larger ones also consume small vertebrates, and a few are partly herbivorous.

Skinks may be either oviparous or ovoviviparous, and a few, such as the European skink, *Chalcides ocellatus*, approach true viviparity, with placenta formation. *Eumeces fasciatus* of the eastern United States is a typical oviparous skink. In Maryland, courting and copulation take place shortly after emergence from hibernation, generally during early May. The eggs are laid some 6 or 7 weeks afterwards. The clutches range from 2 to 18 in number, the smaller clutches being laid by the smaller females. There is also some indication that the size of the clutch decreases toward the northern part of the range. The eggs are deposited in rotten wood or loose soil 50 to 75 mm. below the surface and are brooded during the entire incubation period by the mother. Shortly after deposition, the eggs are approximately 13 by 7 mm. in size but, like the eggs of most lizards except the geckos, they increase in size during incubation and shortly before hatching may be as much as 20 mm. long. They hatch 4 to 7 weeks after deposition. At the time of hatching the young measure 24 to 28 mm. in head and body length.

The Australian Blue-tongued Lizard, *Tiliqua scincoides*, is an example of a live-bearing skink. The young are born in the middle of the summer (January); litters of from 5 to 18 have been recorded. The little lizard is still wrapped in the egg membrane, but in a few seconds it breaks through and immediately devours its own membrane. The young at birth are from 130 to 152 mm. in total length.

Family Anelytropsidae

The only genus in this family is a small (200 mm.) worm-like, limbless lizard, presumably derived from the Scincidae, but differing in the absence of a pectoral girdle and the loss of the temporal arch from the skull. The single species, *Anelytropsis papillosus,* occurs in central and eastern Mexico. Only three or four specimens have been collected, one from under a rotten log near an ant nest and another from under a rock. Of its way of life, nothing is known.

Family Feylinidae

This family was erected for a genus of small, worm-like, limbless scincomorph. Like *Anelytropsis,* members of the genus lack skull arches, but they retain a vestige of the pectoral girdle and are somewhat larger, being about 300 mm. long. Chevron bones are associated with the caudal vertebrae. The four species of *Feylinia* are found only in equatorial Africa. They feed almost exclusively on termites and are most often taken under rotten logs where these insects abound. The life history is unknown.

Family Cordylidae

This small family of scincomorph lizards is somewhat intermediate between the Scincidae and the next family described—the Lacertidae. It includes genera that were formerly classified in two separate families— Cordylidae (or Zonuridae) and Gerrhosauridae. These genera are quite similar, both in structure and in geographic distribution, and it seems best to put them together in a single family. These again are small lizards, the largest only a little more than 600 mm. long, with the tail making up most of the length. Osteoderms are present on the head and body, the limbs are sometimes reduced to stumps, and femoral pores are well developed. The tongue has papillae, but is only feebly nicked anteriorly. The 10 genera are restricted to Africa south of the Sahara and to Madagascar. Both oviparous and ovoviviparous cordylids are known. *Gerrhosaurus v. validus* of southern Africa lays about four soft-shelled eggs in September or October, whereas *Cordylus cordylus* of the same region bears one or two living young.

Family Lacertidae

The members of this family are mostly small, agile, long-tailed lizards. Alone among the Scincomorpha, they show no tendency toward reduction of the limbs. The dorsal temporal arch is complete, but bony dermal plates roof over the temporal fenestra and fuse with the cranial bones when in contact with them, thus obscuring the structure of the dorsal arch. Osteoderms are lacking on the body, though present on the head. The lateral teeth are often bicuspid or tricuspid. The tongue is moderately elongated, deeply notched anteriorly, and covered with scale-like papillae or with transverse plicae. Femoral glands are usually present. Some forms have developed windows in the lower eyelids. The largest species is the Jeweled Lacerta (*Lacerta lepida*) which reaches a length of 750 mm.

The Lacertidae are Old World lizards, occurring in Europe, Asia, and Africa, but not in Madagascar or in the Australian Region. They are most abundant in Africa, comparatively rare in the Oriental Region. The Common Lizard of Europe, *Lacerta vivipara*, is found above the Arctic Circle, farther north than any other lizard. The family includes about 20 genera.

Lacertids are predominantly terrestrial, often living in grassy or sandy places. Some, such as the European Wall Lizard (*Lacerta muralis*), are agile climbers. They are carnivorous, feeding chiefly on insects and other small invertebrates.

Lacerta vivipara, as its name indicates, usually bears living young, but the other lacertids are oviparous. The English Sand Lizard, *Lacerta agilis agilis*, breeds during May and early June. The same male and female will mate together many times. The eggs are laid in June and July and vary from 6 to 13 in number, depending in part on the size of the female. When first laid, they are 12 to 15 mm. long and 8 to 9 mm. wide; shortly before hatching they are from 15 to 20 mm. long. They are hidden under stones or in shallow holes dug in the earth and covered over by the mother. They hatch in 7 to 12 weeks and the young, at hatching, are 56 to 63 mm. in head and body length.

Family Teiidae

This family occupies in the New World a place similar to that filled in the Old World by the closely related Lacertidae. The largest of the teiids (*Tupinambis*) is about 900 mm. long, but most are small. Some are degenerate burrowers with reduced limbs. Osteoderms are absent on the head as well as on the body. The skull arches are present, and the temporal fenestra are open. The tail is quite long. The tongue, like that of the Lacertidae, is long and narrow, deeply forked anteriorly, and covered with papillae. The front teeth are always conical but the lateral teeth on both jaws may be conical, bicuspid, tricuspid, molariform, or even enormous, oval crushers as in the snail-eating Caiman Lizard, *Dracaena guianensis*.

FIG. 15-7. The snail-eating Caiman Lizard, *Dracaena guianensis.*

The family is found only in the western hemisphere, where it is represented by 40 genera. Practically all are restricted to South America; only one, *Cnemidophorus*, reaches the United States.

Most teiids are either terrestrial or fossorial but *Dracaena* is semiaquatic. The larger species are diurnal, but some of the smaller ones are nocturnal. Most, but not all, are carnivorous.

Life history notes are fragmentary, but observations made on several different species seem to indicate that on the whole the teiids have rather uniform life histories. A female of the Coastal Whiptail, *Cnemidophorus tigris multiscutatus*, is known to have mated on May 23 and again on May 25. About 3 weeks later, on June 13, she laid 3 eggs. She mated again on July 3 and on July 22 she again laid 3 eggs. They were immaculate white and ranged between 19.5 to 20.8 mm. in length and 10.9 to 11.0 mm. in width. They increased about 8 percent in length and 35 percent in width during the incubation period of about 80 days. The young were between 110 and 118 mm. long at hatching.

INFRAORDER ANGUINOMORPHA

This infraorder includes a heterogeneous assemblage of animals ranging from the bulky, 10-foot-long Komodo Dragon, *Varanus komodoensis*, to the worm-like Legless Lizard, *Anniella pulchra*, no bigger than a lead pencil. At first glance, they seem to have nothing in common, but underlying structural similarities justify their being placed together. The tongue is divided into two parts, with a notched, inelastic forepart set off by a transverse fold from the elastic hind part, which serves as a sheath when the tongue is withdrawn. The teeth are nearly solid, not hollowed out at the base as are those of other lizards, and are replaced alternately; that is, the new tooth comes up behind, not beneath the older tooth. There are relatively few anguinomorphs living today, but the group includes a number of large, heavily armored, extinct forms, notably the huge, aquatic mososaurs, the most spectacular lizards that ever lived.

The infraorder is divided into two superfamilies, each having three families.

Superfamily Diploglossa

These are moderate-sized to small insect-eating lizards, generally more or less armored with osteoderms. Many have reduced limbs and snake-like bodies. The external naris is not prolonged backward as a slit but is a round or oval foramen. The toothed maxillary extends far back beneath the orbit and the bones of the lower jaw are rigidly joined together. The tail can be regenerated.

Family Anguinidae. This family includes some lizards with well-developed limbs, such as the Alligator Lizards (*Gerrhonotus*), and some that lack limbs, such as the Glass Lizards (*Ophisaurus*) and Slow Worms (*Anguis*). The tropical American genera known as Galliwasps show intermediate stages in limb reduction. The temporal arches are present, the temporal openings are long and narrow and in some forms are roofed over. Palatal teeth may be present or absent. Osteoderms are well developed. Most anguinids are moderate in size. The largest, *Ophisaurus apodus*, may attain lengths of nearly 1,200 mm.

This family includes about 10 genera, most of them found in tropical America. Only two are present in the Old World. *Ophisaurus* occurs in Asia, southeastern Europe, and northern Africa, as well as North America; *Anguis* is found only in Europe, northern Africa, and western Asia.

Most of the anguinids are terrestrial; a few show arboreal tendencies. Some of the species remain hidden in burrows during the day, coming out at night to hunt for insects and other small invertebrates.

Both oviparous and ovoviviparous forms are known, not only within the family, but within a single genus. The San Francisco Alligator Lizard, *Gerrhonotus coeruleus coeruleus*, which is ovoviviparous, copulates in April. The copulatory process is lengthy, sometimes lasting for many hours. The number of young varies from 2 to 15, but usually about 7 are born in late August or September. When first born, they have a snout-to-vent length of 25 to 30 mm. On the other hand, the Oregon Alligator Lizard, *Gerrhonotus multicarinatus scincicauda,* is oviparous. Mating occurs from the middle of May to the middle of June. The eggs, which may number more than a dozen in a single clutch, are laid in late July and early August, in burrows dug by mammals. The young hatch out in September.

Family Anniellidae. This family is represented solely by *Anniella,* the little, shovel-snouted, Legless Lizard of California and Baja California. The bones of the skull are closely united, the temporal arches are absent, the snout is short, and the braincase region is expanded. Palatal teeth are lacking and the osteoderms are reduced. This little lizard is about 250 mm. long, lacks external ear openings, but has lidded eyes. The body is covered on all sides with smooth, rounded, uniform overlapping scales like those of a skink.

Legless lizards lead almost exclusively underground lives but are sometimes found on the surface of the sand under rocks and logs.

Anniella is ovoviviparous; one to four young are born in late summer and fall.

Family Xenosauridae. Two poorly known genera, widely separated geographically and very limited in distribution, are included in this family. *Xenosaurus* is found from southern Mexico to Guatemala; *Shinisaurus* occurs in southern China. They are moderate in size (about 250–375 mm.), with normally developed limbs. Two clearly defined, longitudinal crests, formed by series of enlarged scales, run down the midline of the back. The rest of the dorsal surface is covered with a scattering of smaller scales interspersed with minute granules. The temporal arches are strongly developed and the temporal openings are large and not roofed over by skull bones. The bones of the skull are roughened by the fusion to them of the cranial osteoderms.

Shinisaurus is said to live along streams and feed partly on tadpoles and fish. Beyond this nothing is known of the habits and life histories of these lizards.

Superfamily Platynota

These are moderate-sized to very large predaceous lizards, with jaws adapted more for grasping large prey than for chewing small invertebrates. In line with this, the bones of the mandible are rather loosely joined and there is a tendency toward the development of a hinge within the jaw. The maxillary barely reaches the level of the orbit, so that the marginal teeth are all in front of the eye. The slit-like external nares extend far back in the skull. The tail is not autotomous and the limbs are never reduced. Three families are included in the superfamily, each with but a single genus.

Family Helodermatidae. The Gila Monster (*Heloderma suspectum*) and the Mexican Beaded Lizard (*Heloderma horridum*), found only in southwestern United States and Mexico, are the only poisonous lizards known. Unlike those of the poisonous snakes, the venom glands of *Heloderma* are in the lower jaw and are not connected with the teeth. The poison empties into the mouth through a number of ducts opening between the teeth and the lips. Grooves on the teeth help draw it into the wound by capillary action as the lizard hangs on and chews. The bite is painful, but seldom fatal to man. A few palatine and pterygoid teeth are present; the temporal arch is absent; there are eight cervical vertebrae. The back and the outer surfaces of the limbs are covered with large osteoderms.

These are heavy-bodied, short-tailed, clumsy looking lizards, gaudily marked with dark reticulations on a yellow or orange background or vice versa. The Beaded Lizards may reach a maximum length of 900 mm. The Gila Monster is smaller, being at most about 500 mm. long. In captivity both feed entirely on hens' eggs so it seems probable that in the wild state eggs and nestling birds form a major part of their diet. Poison would seem to be little needed against such prey, and the function of the venom of these animals remains obscure.

Heloderma suspectum has been seen in copulation in mid-July. What

FIG. 15-8. The Gila Monster, *Heloderma suspectum*. It and its relative, *H. horridum*, are the world's only poisonous lizards.

little evidence there is indicates that three to seven eggs are laid during a
period from late July to mid-August. They are buried to a depth of about
125 mm. in an open place, exposed to the sun but generally near a stream
or dry wash. The thin-shelled, white, rather rough eggs are 67 to 75 mm.
long and 33 to 39 mm. wide. The incubation period is apparently 28 to 30
days.

Family Varanidae. All living members of this family are included in a
single genus, *Varanus*, the Monitor Lizards. The smallest monitor, *Varanus
brevicauda* of Australia, is only about 200 mm. long, but most are large, so
much so that they have been mistaken for crocodiles. Indeed, the native
name for the Komodo Dragon, "buaja darat," means land crocodile. This
monitor may reach a length of 300 cm. Monitors lack the venom apparatus
of the helodermatids. The dorsal temporal arch is complete, palatine and
pterygoid teeth are absent, and the osteoderms are reduced or sometimes
lacking. The tail is long and muscular and, like that of the crocodile, is a
formidable weapon. There are nine cervical vertebrae. A monitor holding its
head erect on its long neck has a very alert appearance.

About 30 species of *Varanus* are known. They occur in southern Asia,
Africa, the East Indies, and Australia. All are carnivorous, and the larger
forms are said to be capable of tackling such prey as pigs and small deer.
Although largely terrestrial, they are surprisingly agile climbers for such
bulky animals. Many are quite aquatic and have been seen swimming far
out to sea. This undoubtedly explains their distribution throughout the East
Indies.

The monitors are oviparous. In Thailand, the Common Water Monitor,
Varanus salvator, lays 15 to 30 eggs at the beginning of the rainy season in
June. They are deposited in holes in riverbanks or perhaps in trees beside
the water. They measure about 70 by 40 mm., have rather soft shells, and
are said to taste like turtle eggs.

Family Lanthanotidae. The sole representative of this family, *Lanthano-
tus borneensis*, is known only from along a single river in Borneo. One of the
rarest of lizards, at least in museum collections, it looks like a diminutive
crocodile about 400 mm. long. The back and tail are protected by ridges of
raised tubercles. The dorsal temporal arch is lacking, a few teeth are present
on the palatine and pterygoid bones, and there are nine cervical vertebrae.
These animals bear a superficial resemblance to *Heloderma* and were long
classified with this genus, but they have no venom apparatus and are
anatomically much closer to *Varanus*. Nothing is known of their life history
and habits.

INFRAORDER ANNULATA

Included within this infraorder are a number of highly specialized little burrowers that have departed so far from the normal lacertilian type that some herpetologists feel they should be placed in a separate suborder. Their bodies are elongate and nearly uniform in diameter, and their tails are short.

Except for one genus, *Bipes*, which has short front legs, external limbs are lacking and the girdles are vestigial. There are no external ear openings and the eyes of the adults are hidden under the skin. Osteoderms are absent. The soft skin of the body is folded into numerous rings divided into quadrangular areas representing the flattened and reduced scales. These rings, combined with the absence of limbs and the cylindrical body form, give the Annulata a remarkable resemblance to earthworms. They are all included in a single family.

Family Amphisbaenidae

The skull of an amphisbaenid is highly specialized for digging. The bones are closely united, the snout region is short and generally expanded, the braincase almost completely enclosed. The skull arches are lost. The teeth are large but few in number, and are absent from the palate. Only one functional lung, the left, is present. Most of the species are about 300 mm. long. The largest, *Monopeltis* of Central Africa, attains lengths of about 675 mm.

The family contains more than 20 genera and about 100 species. Members are found in Africa and the Mediterranean countries, in South America and

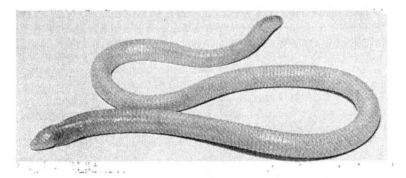

FIG. 15-9. The Florida Worm Lizard, *Rhineura floridana*, locally known as "Graveyard Snake."

the West Indies, and range north through Mexico to Baja California and possibly to Arizona. One species, *Rhineura floridana,* is isolated in Florida.

The name of the family comes from two Greek words which, translated literally, mean "walk at both ends," in reference to the ability of amphisbaenids to move backward as well as forward in their underground tunnels. A most appropriate common name is Worm Lizard. In Florida, where they are best known to grave diggers who turn them up while digging in the sandy soil of cemeteries, they are sometimes called Graveyard Snakes. They feed on worms and small insects, especially ants and termites. So far as is known, most amphisbaenids lay eggs, but the North African *Trogonophis* gives birth to living young.

INCERTAE SEDIS

The words *"incertae sedis"* (of uncertain place) appearing in a classification, mean, in plain English, that the author does not know enough about the relationships of a group to decide where it should be placed. We list as *incertae sedis* two families, the Night Lizards, Family Xantusiidae, and the Blind Snakes, Family Typhlopidae.

Family Xantusiidae

The xantusids resemble the geckos in the shape of the vertebrae and in the structure of the eye, which has an elliptical pupil and fused eyelids, the lower enlarged into a transparent window. On the other hand, they are like the Scincomorpha in having well-developed temporal arches, with the supratemporal fossa roofed over by the parietal. Post-cloacal sacs and bones have been reported for one species but others apparently lack them.

The family contains only four genera; *Xantusia* is found in southwestern United States, *Gaigeia* and *Lepidophyma* in central Mexico and Central America, and *Cricosaura* in Cuba.

These are little lizards, less than 150 mm. long, with well-developed legs and tails. The dorsal scales are granular, the belly scales rectangular plates. The tail is easily shed. Secretive and nocturnal, they are most often found, in the United States at least, hiding under the flakes of rock that form on the faces of granite boulders in desert regions. The few observations that have been made indicate that they are insectivorous.

Xantusids are truly viviparous. Mating of *Xantusia vigilis,* the Desert Night Lizard, takes place in May, and ovulation occurs one to four weeks later. Usually two eggs are formed at one time, another way in which the Night Lizards resemble the geckos, but rarely only one embryo is found, usually in the right oviduct. Occasionally, three embryos are present. Early in embryonic development, a simple, cellophane-like shell is formed, but it soon disintegrates, and a chorio-allantoic placenta develops. Gestation takes about three months.

Surprisingly enough, these lizards have developed the typically mammalian custom of eating the fetal membranes. The membrane ruptures before the young is born and remains in the cloaca. The female grasps the protruding edge in her mouth, gradually draws it out, and swallows it.

Family Typhlopidae

These highly aberrant little animals have been classified both as lizards and as snakes, and it has also been suggested that they should stand alone, as a separate suborder of the Squamata. Recent anatomical studies indicate that they are probably more closely allied to the Lacertilia than they are to the Serpentes. They are worm-like burrowing creatures with cylindrical bodies and short tails. The largest species may be 750 mm. long but most are less than 200 mm. The eyes are more or less distinct but are covered by the head shields. The transversely placed maxilla is loosely attached to the skull and bears teeth that are directed backward. The premaxillary, palatine, and pterygoid bones lack teeth. Some species have a single tooth at the tip of each mandible, but there is never a row of teeth on the lower jaw. Zygosphenes and zygantra are present on the vertebrae. The pectoral girdle is absent. The pelvic girdle may be represented by pubic, ischial, and iliac elements, with traces of pubic and ischial symphyses, or by a single, rodlike bone on each side, or it may be absent entirely.

There are five genera in the family. Four are confined to Central and South America, and the fifth, *Typhlops,* is widely distributed. It is found in

FIG. 15-10. The Dominican Blind Lizards, *Typhlops dominicana.* For many years considered snakes, these creatures are now grouped with the lizards.

Central and South America, the West Indies, southern Europe, Africa, southern Asia, the East Indies, and Australia.

Very little is known of the breeding habits of these small and secretive lizards. The Brahminy Blind "Snake," *Typhlops braminus*, lays 2 to 7 tiny, elongate eggs, each about 12 mm. long and 4 mm. in diameter. But apparently not all typhlopids are oviparous, for a specimen of *Typhlops diardi* was found to contain 14 perfectly developed embryos.

Collateral Reading and General Reference

Boulenger, G. A. *Catalogue of the Lizards in the British Museum (Natural History)*. London: British Museum, 1885–1887.

Camp, C. L. "Classification of the Lizards." *Bulletin of the American Museum of Natural History*, vol. 48, article 11, 1923. (A sound, basic work in lizard classification.)

Cope, E. D. The Crocodilians, Lizards and Snakes of North America. Report United States National Museum for 1898. Washington: Smithsonian Institution, 1900. (A basic work on North American forms.)

McDowell, S. B., Jr. and C. M. Bogert. "The systematic position of *Lanthonotus* and the affinities of the Anguinomorphan lizards." *Bulletin of the American Museum of Natural History*, vol. 105, article 1, 1954. (A modern treatment of one infraorder.)

Smith, H. M. *Handbook of Lizards*. Ithaca: Comstock, 1946. (A modern handbook, restricted to North American forms.)

Smith, M. A. *Fauna of British India, Reptilia and Amphibia*. Vol. II. Sauria. London: Taylor and Francis, 1935.

Underwood, G. "On the classification and evolution of geckos." *Proceedings of the Zoological Society of London*, vol. 124, part 3, 1954. (A modern survey of a major group of lizards.)

SNAKES

AN INTEREST in snakes has probably led more people to study herpetology than an interest in all the other groups of herptiles combined. While it is true that many of these students have later turned to more prosaic forms, such as the salamanders or turtles, it is still the snakes that first attracted their attention.

The approximately 2,700 kinds of snakes in the world form the suborder Serpentes of the order Squamata. Like the lizards in the suborder Lacertilia, they have undergone extensive adaptive radiation and have come to occupy most of the major habitats of the world. Some are burrowers; these are mostly small in size and their eyes are hidden beneath the scales of the head. Others have taken to the trees and seldom come to the ground. One group is entirely marine; these include the only modern reptiles to abandon the land completely, for even the sea turtles come to shore to lay their eggs. In size the snakes range from tiny forms only about 100 mm. long to the Anaconda and Reticulated Python which reach lengths of more than 9 m.

Snakes are not as important economically as turtles. Some, probably most, are edible, but they are seldom utilized as food although in the Orient not only the Pythons but also the poisonous sea snakes are esteemed by some peoples. Their skins make a leather suitable for fancy belts, pocketbooks, and shoes, but one that has never been much used for a wider range of articles. Structurally, the snakes are so highly modified that they are not popular specimens in anatomical laboratories, and indeed there is no good description of the anatomy of a snake comparable to those available for the salamander and the frog. Their major contribution to mankind is the control of pests, particularly the destructive rats and mice.

It would seem that, with all the interest attached to the snakes, the major problems of their classification and evolutionary relationships would long ago have been worked out. Many of the genera and species have been analyzed in great detail, and the analyses have become valuable tools for zoogeographic and evolutionary studies. But organization of the snakes into higher categories remains one of the major problems of herpetology.

Snakes are elongate animals with no girdles or limbs, or occasionally with vestigial pelvic girdles and hind limbs. They lack a sternum, external ear opening, tympanic membrane, middle ear, and eustachian tube. Except in some burrowing forms, the immovably fused and transparent eyelids form a protective window, the brille, beneath which the eye moves. The viscera are elongated, and the left lung is smaller than the right or altogether absent. The tongue is long, forked, and protractile. There is no urinary bladder. Like the lizard, the snake's body is covered with scales, the vent is transverse, and the copulatory organs are paired.

The skull of a snake is more specialized than that of a lizard. The brain cavity is completely enclosed anteriorly by dermal bones. The higher forms have the bones of the facial region and jaws loosely joined to each other and to the cranium so that they can be spread apart. The two halves of the lower jaw are not fused but are connected by a ligament. Each half of both upper and lower jaw can be moved independently of the other half. This allows a snake to engulf objects that look impossibly large. It seems to "walk" its mouth around its food by a forward movement of first one side of the mouth and then the other.

The elongation of the vertebral column results from an increase in the number of vertebrae rather than from a lengthening of individual vertebrae.

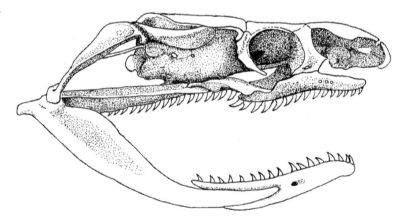

FIG. 16-1. The skull of a colubrid snake, *Drymarchon corais couperi*, the Indigo Snake of the southeastern United States.

The earliest reptiles seem to have had about 25 presacral vertebrae. Because of the absence of the girdles, it is difficult to classify the snake vertebral column into regions. The total number of vertebrae reported for different species ranges from 141 to 435. Most of these are dorsals. The caudals (i.e., those located behind the region of the vent) average about 50 to 60 in number, although in the short-tailed burrowing forms the number is much lower and in some long-tailed snakes the ratio of dorsals to caudals may be only 1:1.

The vertebrae are complicated in structure. In addition to the usual articulating facets—the prezygapophyses and postzygapophyses—each has a pair of zygantra and a pair of zygosphenes (see Fig. 3-7). Thus each vertebra has five points of contact with the anterior one: the centrum, the two prezygapophyses which fit against the postzygapophyses, and, dorsal to the zygapophyses on the neural arch, the zygosphenes which articulate with the zygantra. The vertebrae of the anterior part of the trunk have hypapophyses—ventrally directed processes from the centra—to which muscles are attached. In many groups these are present on the posterior dorsals as well. In place of the hypapophyses, the caudal vertebrae frequently have haemopophyses—paired ventral projections from the centra that surround the caudal blood vessels.

The snakes are divided into ten families.

FAMILY BOIDAE

This family includes the largest snakes in the world, the huge Pythons, Boa Constrictors, and Anacondas. (The Boa Constrictor belongs to the genus *Constrictor*. Members of the genus *Boa* are not so large.) Many boids, though, are much smaller than these giants. The Rubber Boa (*Charina*) of the southwestern United States averages about 450 mm. in length. Boas are typically stout-bodied, short-tailed snakes. The palatomaxillary arch is movably attached to the rest of the skull. Teeth are present on the maxillary, palatine, pterygoid, and dentary bones, and sometimes on the premaxillary. The ventral scales form enlarged, transverse plates; the dorsal scales are small and are sometimes iridescent. Except in one subfamily, vestiges of the pelvis and hind limbs are present, the latter terminating in claw-like spurs which are usually visible on either side of the vent and are longer in the male than in the female.

Boids occupy a variety of habitats. Many of the small forms are burrowers in sandy soils. Some are arboreal, with short, more or less prehensile tails. The huge Anaconda of South America (*Eunectes*) is largely aquatic and can remain submerged in water for long periods of time. A number of the large

FIG. 16-2. The largest Old World snake, the Reticulated Python, *Python reticulata*. It is exceeded in size only by the Anaconda of South America, which has a maximum recorded length of 1,143 cm.

boids, such as the Indian Python (*Python molurus*) seem to be equally at home in trees or in water.

The boids feed largely on birds and mammals, and usually kill their prey by constriction. Contrary to popular opinion, they do not crush the bones of their victims. Two or three coils of the snake's body are wrapped around the upper trunk of the prey. These exert enough pressure to stop breathing, and the animal suffocates.

The vestigial hind limbs of the boids are apparently functional structures used in courtship. The male Boa Constrictor vibrates them rapidly and rubs them on the back and flanks of the mate he has chosen. By beating with his spurs, the male Anaconda stimulates the female to move her body forward and place it so that copulation can begin. A Python, after having arranged himself alongside the female and placed the anterior part of his body on her back, taps the region about her cloaca with his claws in a slow and rhythmic manner. This may continue for as long as two hours and stops only when the female inclines her anal region on the side and allows the male to insert his hemipenis in her cloaca.

FIG. 16-3. The external portion of the hind limb of the Cuban Boa, *Epicrates angulifera*. These spurs apparently are functional during courtship.

Both oviparity and ovoviviparity are displayed by the Boidae. The Reticulated Python lays eggs, the number depending partly on the size of the mother. A large, full-grown female may lay as many as 100 eggs. On the other hand, small females have been known to lay as few as 15. The incubation period lasts 60 to 80 days and the newly hatched young are 600 to 750 mm. long. The pythons brood their eggs, the mother coiling around them and perhaps providing extra heat through an elevation of her body temperature. The amount of such an increase is still a matter of debate. Some authors have recorded the body temperature of the brooding animal to be no higher than that of the surrounding environment, while in other species, at other times, body temperatures for the female on the eggs have been reported to be 6 or 7 C. warmer than for the nonbrooding male, and 12 to 15 C. warmer than the air. From such analysis as can be made of these divergent results, it would seem that the body temperature of the female may increase during the brooding process but that the amount of increase varies from one species to another. The state of high temperature in the female is more intense at the beginning of brooding than toward the end of it. Many more observations on temperature and brooding are needed before the overall pattern can be determined.

On the other hand, many of the snakes of this family are ovoviviparous. The Anaconda (*Eunectes*) of South America gives birth to 4 to 39 young, each about 800 mm. in length, and the Rosy Boa (*Lichanura*) has been known to give birth to 6 young, each measuring about 280 mm. in length.

The boids are divided into four subfamilies.

Subfamily Boinae

The great majority of the boas and pythons (about a dozen genera) are placed in this subfamily. It is customary to speak of the New World forms as boas and those of the Old World as pythons, but the two groups are obviously closely related and should not be placed in separate subfamilies simply because they are separated geographically. All forms have hypapophyses only on the anterior dorsals. The left lung is well developed, though smaller than the right. This is essentially a tropicopolitan group, found in warm countries throughout the world.

Subfamily Tropidophinae

This subfamily includes only two genera, *Trachyboa* and *Tropidophis*, which are small snakes of the West Indies and northern South America. Hypapophyses are present on all the dorsal vertebrae. These snakes have only one true lung, the right, but a well-developed, lung-like structure, the tracheal lung, is also present on the dorsal wall of the trachea.

Subfamily Bolyeriinae

Two genera, *Bolyeria* and *Casarea*, found only on Round Island (a small island off Mauritius, near Madagascar) make up this subfamily. They are rather small, semifossorial snakes, differing from all other boids in that the

FIG. 16-4. The Cuban Boa, *Epicrates augulifera*, in its characteristic feeding posture.

pelvic girdle and hind limb rudiments are completely absent. Hypapophyses are present on all dorsal vertebrae and both lungs are well developed.

Subfamily Erycinae

This subfamily includes four genera of small boids. *Charina* (Rubber Boa) and *Lichanura* (Rosy Boa) are found in western North America, *Eryx* (Sand Boa) in Africa and Asia, and *Engyrus* (Pacific Boa) in the East Indies. They are short-tailed, small-eyed, mainly fossorial snakes which resemble the Boinae but lack zygapophyses on the caudal vertebrae. Apparently all the Erycinae are ovoviviparous.

FAMILY ANILIIDAE

These are stout-bodied, short-tailed, cylindrical snakes which may be as long as 900 mm. The scales are small and smooth, those on the ventral side being slightly enlarged. The bones of the skull are solidly united. Teeth are present on the maxillary, palatine, and pterygoid bones, on the dentary bone of the lower jaw, and, in one genus (*Anilius*), on the premaxillary as well. Hypapophyses and haemopophyses are absent. A vestigial pelvis and rudimentary hind limbs are present, the latter projecting as claw-like spurs on either side of the vent.

The family includes three genera of burrowing snakes: *Anilius*, the beautiful red and black False Coral Snake of northern South America; *Anomochilus* of Sumatra; and *Cylindrophis*, the Pipe Snake or Two-headed Snake of southeastern Asia.

Snakes in which the bones of the skull are solidly fused cannot open their mouths as widely as can most snakes and hence are largely restricted to small food items like insects and worms. However, *Cylindrophis rufus* is reported to feed on other snakes and eels and to be able to dispose of a meal even longer than itself. The Aniliidae are ovoviviparous. *Cylindrophis maculatus* produces two or three young that may be more than 125 mm. long at birth.

FAMILY UROPELTIDAE

The Uropeltidae (Rough-tails) are secretive, frequently fossorial snakes with rigid, cylindrical bodies and very short tails. Most are less than 600 mm. long. Many species are brightly marked with red, orange, or yellow; some are a shiny, iridescent black. The pupil of the eye is round. The ventral scales are but little larger than the dorsals. The bones of the skull are more solidly united than are those of any other family of snakes. Lacking feet, burrowing snakes must dig with their heads; the remarkably solid skull of the Rough-tails is probably an adaptation to this mode of life. The maxilla has 6 to 8 teeth, the mandible 8 to 10, the palatine 3 or 4 minute ones or none at all. The occipital condyle projects markedly beyond the back of the skull. There are no vestiges of hind limbs or a pelvic girdle.

The most striking characteristic of the Rough-tails is the enlarged scale at the end of the tail. It is either very rugose, or spiny, or reduced to two short points. Freshly caught specimens often have the tail coated with mud. The purpose of this unique appendage has never been satisfactorily explained.

There are seven genera in the family. All are found in damp places in mountainous regions of southern India and Ceylon. These snakes are quiet and inoffensive; they do not bite when handled, nor do they apparently show any fear. When picked up they do not try to escape but will entwine themselves around the fingers or a stick and remain in that position for long periods of time. They are easily kept in captivity and have been known to eat immediately after being caught. Like most small burrowing snakes, they feed on worms and soft-bodied insect larvae. So far as is known, all are ovoviviparous, producing from three to eight young at a time.

FAMILY XENOPELTIDAE

This small family contains only one species, *Xenopeltis unicolor*, the Sunbeam Snake of southeastern Asia. Its common name comes from the highly iridescent scales. As it crawls along in sunlight it flashes with electric blue, emerald green, blood red, purple, and copper. This brilliant display is seldom seen, however, for these snakes are secretive and largely nocturnal. The body is cylindrical, the tail short, and the ventral scales are enlarged to form transverse plates. The female may reach a length of more than 900 mm., but the male is somewhat smaller. The bones of the skull are united. The small teeth are set close together and are strongly curved; there are 4 or 5 on each side of the premaxilla and 35 to 45 on each maxilla. The palatine, pterygoid, and dentary also bear teeth. Hypapophyses are absent on the posterior dorsal vertebrae. There is no trace of the pelvic girdle or hind limbs. *Xenopeltis* has two well-developed lungs, the left one being about half as large as the right.

Sunbeam snakes are frequently found beneath logs and stones in rice fields and gardens near human habitations. They can burrow rapidly in soft earth and those kept in captivity usually spend the day under cover and come forth only at night. They feed on other snakes, small rodents, and frogs. Apparently nothing has been recorded about their breeding habits.

FAMILY LEPTOTYPHLOPIDAE

The Slender Blind Snakes or Thread Snakes are small, degenerate, burrowing forms that bear a close superficial resemblance to the members of the Typhlopidae (now placed with the lizards) but differ from them in many structural features. No teeth are present on the upper jaw or roof of the mouth and the maxilla borders the mouth instead of being placed transversely. Rows of teeth appear on the mandible. The pelvis consists of the ilium, ischium, and pubis, but is not attached to the vertebral column. A vestigial femur is usually present and may project through the skin in the anal region. The cylindrical body is covered with uniform scales; the eyes lack brilles but are covered by the head shields. The largest species is only about 300 mm. long; the smaller forms are little more than 100 mm. long.

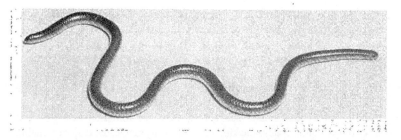

FIG. 16-5. The Texas Blind Snake, *Leptotyphlops dulcis.* In the Old World, members of the genus *Leptotyphlops* are often called Thread Snakes.

The species in this family are all included in a single, widely distributed genus, *Leptotyphlops,* which is found in Africa, southwestern Asia, southwestern United States, and tropical America.

As for many other subterranean creatures, the life history of the Thread Snakes is poorly known. They live beneath the surface of the ground, but may come out during the early evening hours to wander about for a short time. They feed largely on termites, adroitly sucking out the contents of the abdomen. The eggs are long and slender and number about four.

FAMILY ACROCHORDIDAE

The weird-looking Wart Snakes are blunt-headed, small-eyed, ungainly creatures with unusually stout bodies for snakes. The largest females may

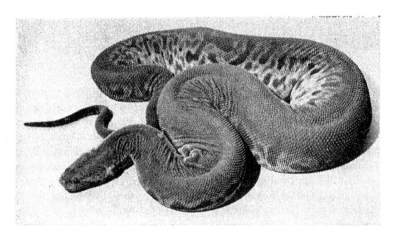

FIG. 16-6. *Acrochordus javanicus,* the Javanese Wart Snake.

reach lengths of 180 cm. The skin is loose and the head and body are covered with small, granular or tuberculate, juxtaposed scales. The ventral scales are not enlarged, and those of the head are minute and sometimes pointed in the region of the nostrils. This is the source of the common name, Wart Snake. (They are also known as Elephant's Trunk Snakes.) Hypapophyses are well developed on all the trunk vertebrae. Pelvic girdle and hind limbs are lacking. The tail is short and compressed.

Wart Snakes are aquatic fish eaters; they are found in estuaries and enter the sea quite freely. Lacking enlarged ventral shields, they are unable to glide normally on land, but progress by a slow, clumsy, heaving of the body. Since they are ovoviviparous, they do not need to come ashore to lay eggs. One female has been reported as giving birth to 27 young. The family includes only two genera, *Acrochordus* and *Chersydrus,* which are found in India, Indo-China, and the Indo-Australian archipelago.

FAMILY COLUBRIDAE

This huge, cosmopolitan family includes the great majority of the snakes of the world. Its members occupy a wide variety of habitats; some are terrestrial, some aquatic, some arboreal, and a few fossorial. Most are moderate to small in size, and none attains the bulk of the great boas. Almost all are harmless to man.

All colubrids lack rudiments of the pelvic girdle and hind limbs, and the left lung has disappeared. The facial bones are movable and are loosely attached to the skull (see Fig. 16-1). Usually the belly scales are as wide as the body. Teeth are normally present on the maxillary, palatine, ptery-goid, and dentary, but are never found on the premaxillary. Most species have solid teeth, without grooves (aglyphous), and unconnected with any poison glands. A few have several of the rear teeth grooved (opisthogly-phous). The supralabial gland above is specialized to produce a poison which is channeled down the grooves. Such rear-fanged snakes do not in-ject poison by striking, but by chewing an object that has been taken into the mouth. The poison is thus used, not for capturing prey, but for quieting the struggles of the animal being swallowed, and for initiating digestion. Most rear-fanged snakes are small and harmless, but both the African Boom-slang (*Dispholidus*) and the African Vine Snake (*Thelotornis*) have caused human fatalities, and several other species may also be dangerous.

As would be expected in such a large and varied family, colubrids show great diversity in feeding habits. Some will eat almost anything they are able to catch and engulf, whereas others, such as the egg-eating and snail-eating snakes, have specialized diets. The smaller forms eat worms and insects, and many of the larger ones feed exclusively on birds and mammals and usually kill their prey by constriction. Aquatic colubrids prey on fish

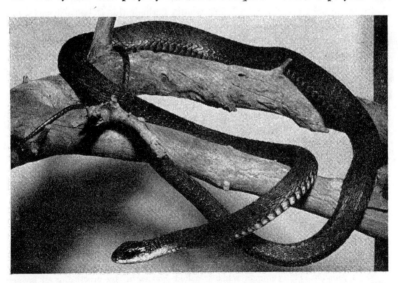

FIG. 16-7. The African Boomslang, *Dispholidus typus*, a snake potentially fatal to man.

and amphibians. The King Snakes (*Lampropeltis*) seem to be especially fond of other snakes.

Oviparous, ovoviviparous, and viviparous forms are present in the Colubridae. The life history of the Racer (*Coluber constrictor*), of the eastern United States, is typical of the oviparous members of the family. Mating has been observed in May, and the eggs are laid in decaying vegetable matter through June and early July. There may be as many as 25 eggs in a clutch, but the average is probably about 12. Eggs are about 50 mm. in length, and are somewhat elongated and granular in texture. The young, which hatch in August, are between 200 and 300 mm. long.

The breeding habits of the Northern Water Snake, *Natrix sipedon sipedon*, may be taken as normal for the ovoviviparous forms. Mating takes place in the early spring and the young are born in the late summer and early fall. The average number reported in 8 broods was 31, with the number per brood ranging from 16 to 40, but much larger broods have been recorded, some snakes giving birth to more than 75 at one time. Length of the young averages about 225 mm. at birth.

True viviparity, with placenta formation, occurs in the Garter Snake, *Thamnophis sirtalis*.

If any generalization can be made about the mode of life history of the Colubridae it is that the more terrestrial forms lay eggs whereas the more aquatic ones give birth to living young. Perhaps this is because eggs laid in the places normally inhabited by aquatic snakes might be endangered by flooding. But even this generalization cannot be carried too far, for the highly aquatic Mud Snakes of the genus *Farancia* lay 50 to 100 eggs which the female broods. There is variation even within a single genus; the young of American forms of *Natrix* are born alive, whereas females of the European *Natrix* (less aquatic than their American congeners) lay eggs.

FIG. 16-8. The Northern Water Snake of the eastern United States, *Natrix s. sipedon*. This common snake is called "moccasin" by practically all laymen in its range.

Such a large and unwieldly family as the Colubridae can be much more easily discussed if it is subdivided into smaller groups, or subfamilies, but this is very difficult to do. An adequate classification of these snakes remains one of the most pressing problems in herpetology. The family is here divided into nine subfamilies. Seven small groups of specialized forms have been separated from the others. The remainder of this bewildering assemblage has been placed in two large, heterogeneous, rather ill-defined subfamilies, depending on whether the posterior hypapophyses are present (Natricinae) or absent (Colubrinae). This leaves unresolved the problem of certain genera (*Boiga, Chrysopelea*) in which posterior hypapophyses are present in some species, absent in others. It must be emphasized that this classification is only provisional, and will probably be greatly modified as our knowledge of the anatomy of the colubrid snakes increases.

Subfamily Xenoderminae

This subfamily includes four small genera of little snakes: *Xenodermus* and *Stoliczkia* of southeastern Asia and the East Indies, *Cercaspis* of Ceylon, and *Xenopholis* of central South America. Each contains only one species except *Stoliczkia*, which has two. They are classified together because of the peculiar structure of their vertebrae. The neural spine is elongated and may be folded over dorsally to form a flat surface. The prezygapophyses are much expanded in the Asiatic forms; the postzygapophyses in the South American genus. The occipital condyles are very small. The small scales of *Xenodermus* and *Stoliczkia* are completely or almost completely attached to the dermis and are more or less separated from one another by bare skin.

Habits of these snakes are little known, but at least one of them, *Xenodermus javanicus*, is nocturnal and lives in loose wet earth. It is frequently found in cultivated fields, and feeds on frogs. Two to four eggs are laid at a time.

Subfamily Homalopsinae

These aquatic snakes have the hypapophyses developed on all dorsal vertebrae. The last two or three maxillary teeth are grooved and usually enlarged. The crescent-shaped nostril is on the upper surface of the snout and has a valve that can be closed when the snake is submerged. The ventral scales are narrow, rather than wide as are those of the aquatic members of the subfamily listed next. These snakes are usually stout of body

FIG. 16-9. The Red-bellied Water Snake, *Natrix erythrogaster*, of the eastern United States in the act of shedding. The snake will ultimately crawl completely out of the shed skin, leaving it more or less entire and inside out.

and short of tail. The largest, *Enhydris bocourti*, seldom exceeds lengths of 90 cm.

The Homalopsinae are equally at home in fresh or salt water and are occasionally found on land in the vicinity of ponds and streams. They feed chiefly on fish, which are often swallowed under water. As usual in aquatic snakes, the young are born alive. The ten genera included in the subfamily range from southeastern Asia, including parts of India and China, throughout the Indo-Australian archipelago to the north coast of Australia. Seven of the 10 genera have only one species each. *Enhydris*, however, includes about 16 nominal forms. *Enhydris tentaculatum* is remarkable for having a pair of mobile, scaled, rostral appendages, one on either side of the snout. The function of these curious structures is unknown.

Subfamily Natricinae

The natricines have well-developed hypapophyses throughout the vertebral column, wide ventral scales, and nonvalvular nostrils. Both aquatic and terrestrial forms are included in this heterogeneous assemblage: some are large, others quite small; some are aglyphous, others opisthoglyphous; some are oviparous, others ovoviviparous or viviparous. This large subfamily occurs in Europe, Asia, Africa, North America, and northern Australia. There are many genera and many species, including the familiar water snakes (*Natrix*) and garter snakes (*Thamnophis*) of the United States. The enormous genus *Natrix*, with about 80 species, is found on every continent except South America.

Subfamily Sibynophinae

These rather small, slender, graceful snakes have well-developed hypapophyses on all dorsal vertebrae and a movable joint in the lower jaw between the dentary and the surangular bones, a character distinguishing them from all other colubrid snakes. The numerous teeth are small and uniform in size. There are only three genera, *Sibynophis* of the Orient, *Parasibynophis* of Madagascar, and *Scaphiodontophis* of Central America. These snakes inhabit the forest floor in mountainous regions, feeding chiefly on lizards and laying from two to four eggs at a time.

Subfamily Colubrinae

This one cosmopolitan subfamily includes the majority of the snakes of the world. The hypapophyses are reduced on the posterior dorsal vertebrae as they are in all of the following subfamilies of the Colubridae. The nostrils are lateral in position; the ventral scales are well developed; the teeth may be solid or the posterior two or three may be grooved. The head is covered with large symmetrical shields.

Most of the colubrine snakes are either terrestrial or arboreal. Besides such familiar types as the racers (*Coluber*) and rat snakes (*Elaphe*) the subfamily includes a number of slim, long-bodied, tree snakes, many of them bright green. The famous "flying snake" (*Chrysopelea*) of the East Indies is an arboreal form that is able to glide through the air. As it launches itself from a branch it straightens out its body and draws in its belly scales to form a concave surface. Colubrines vary greatly in size. *Ptyas mucosa*,

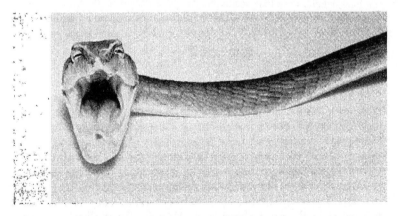

FIG. 16-10. *Dryophis nasutus*, the Indian Green Tree Snake.

FIG. 16-11. The Blunt-headed Tree Snake, *Imantodes cenchoa*,
a Neotropical species.

the Greater Rat Snake of Asia and the East Indies, may be more than 300
cm. long, whereas the Flat-headed Snake (*Tantilla gracilis*) reaches a maximum length of only 225 mm. Most colubrines are oviparous but a few are
ovoviviparous.

Subfamily Pareinae

The Bluntheads are slender little snakes with short, wide heads, slim
necks, and big child-like eyes. Hypapophyses are present only in the cervical
region. The mouth opening extends far back beyond the fringe of the buccal
membrane. The nasal gland (one of the salivary glands) is enormous. The
dentary bone is immovably fused to the surangular. The most striking feature in this subfamily is the arrangement of the scales under the lower jaw.
In most snakes these chin shields are separated at the midline by a furrow,
the mental groove, which is lined with distensible skin. This allows the
two halves of the jaw to be spread widely for the swallowing of large

prey. In the Pareinae, the chin shields of the two sides dovetail and there is no mental groove. Consequently, their jaws cannot be spread widely and their diet is restricted to such small items as snails, slugs, and grubs. They apparently cannot cope with the shell of a snail, but use their sharp teeth to extract the body before swallowing it.

The Pareinae are found in southeast Asia and the East Indies. There are only two genera—*Aplopeltura* with a single species, and *Pareas* comprising about fifteen species. Some are terrestrial, and others are arboreal. They are quiet inoffensive little snakes, mostly nocturnal. So far as is known, they are all oviparous.

Subfamily Dipsadinae

The Slug-eating Snakes are the New World counterparts of the Pareinae, which they resemble in appearance and in habits. The mental groove is present in some, absent in others. The subfamily ranges from Mexico to Brazil and includes three genera (*Dipsas, Sibon,* and *Sibynomorphus*).

Subfamily Dasypeltinae

This and the following subfamily comprise the egg-eating snakes. For an animal not much bigger around than a man's finger to engulf an object the size of a hen's egg is indeed an astounding feat. These snakes accomplish this by one of the most striking series of adaptations to a specialized feeding habit known. There is no mental groove. Instead, the skin along the angle of the mouth and cheek region is especially modified for expansion. The teeth are minute, reduced in number, and restricted to the posterior parts of the maxilla, palatine, and dentary. The upper jaw elements are rigidly fused together, but the bones of the lower jaw are very loosely connected. Most remarkable of all, some of the cervical hypapophyses extend down to pierce the esophagus. Those in front have their ventral edges enlarged into sledlike runners, while those behind are modified to form elongate, forward-pointing spines. The egg, which is swallowed whole, glides down the runners, and is forced against the sharp edges of the posterior hypapophyses thus being cut open. The contents of the egg pass into the stomach and the crushed shell is then regurgitated.

There are about six species of *Dasypeltis* widely distributed in tropical and southern Africa. It is a rather short, slender snake, attaining lengths of about 750 mm. Scales are strongly keeled, those in the vertebral row being about the same size as the other dorsals. *Dasypeltis* is oviparous.

Subfamily Elachistodontinae

This subfamily was erected for a single species of very rare snake—only five specimens have been collected. *Elachistodon* closely resembles *Dasypeltis* in many respects. It lacks the mental groove, and has specially modified, distensible skin at the angle of the mouth and in the cheek. The teeth are restricted to the posterior part of the maxilla, palatine, and dentary, the upper jaw elements are rigidly tied together, and the bones of the lower jaw are loosely articulated. Some of the hypapophyses pierce the esophagus, though they are not as greatly modified in shape as those of *Dasypeltis*. *Elachistodon* differs from *Dasypeltis* in having one or two enlarged and grooved teeth on the rear of the maxilla, and in having smooth scales, with those of the vertebral row enlarged. A large pit is present in the nasal scale. Similar pits in other snakes are known to function as thermo-receptors for locating warm-blooded prey.

Elachistodon westermanni has been found only in northeastern India. Since it resembles *Dasypeltis* so closely, it seems highly probable that it too feeds largely on eggs. The first specimen taken had an amorphous mass in the stomach that may have been egg yolk. The grooved teeth and nasal pits suggest, but do not prove, that it may also feed upon small birds and mammals occasionally. The original specimen had seven hard-shelled eggs in the oviduct, indicating that the species is oviparous.

FAMILY ELAPIDAE

This is the family of the extremely poisonous coral snakes, cobras, mambas, and kraits. Like the members of the two following families, they have venom fangs in the front part of the upper jaw. These three families have given all snakes a bad name, though the poisonous snakes comprise but a small part of the snake fauna of the world. Snakes with fangs of this sort are called proteroglyphs. The fang of the Elapidae is a more or less enlarged, canaliculate tooth, which is held permanently in an erect position and fits into a pocket in the gum tissue on the outside of the mandible but inside the lip when the jaw is closed. The canaliculate tooth has apparently evolved from a grooved tooth like those of the opisthoglyphous colubrids. The groove has sunk in to form a horseshoe-shaped cavity. In the elapids the gap between the ends of the shoe is usually more or less filled in with calcium but it still shows as a furrow on the front surface of the tooth. The duct from the poison gland is not attached directly to the fang but expands into a small cavity in the gum above the opening of the tooth canal. Two fangs are normally present on each maxilla, lying side by side, though usually only one at a time is firmly attached and functional. Each is followed by a series of developing replacement fangs. Snake teeth are constantly being shed and replaced and this arrangement insures that the snake is never without functional fangs. When the fang on the inner side of the maxilla drops off, the one on the outer side is either ready to be used or is already in use. It serves while the next replacement fang on the inner side is growing into place. The maxillary bone is shortened and probably represents ·only the hind part of the ·maxilla found in the rear-fanged, opisthoglyphous snakes. It usually bears one or more small, solid or slightly grooved, teeth behind the fang. Teeth are also present on the pterygoids, palatines, and dentaries. The facial bones are movable. Hypapophyses are developed throughout the vertebral column, and the pelvic girdle and left lung are lacking.

The Elapidae are found throughout the tropical and subtropical regions of the world, but are most numerous in Australia, where most of the snakes belong to this family. They are absent from Europe today but fossil forms have been described from the Miocene and Pliocene of France. About 30 genera are known, including the longest of all poisonous snakes, the King Cobra (*Ophiophagus hannah*) which may be more than 540 cm. long.

The snake so often pictured with Indian snake charmers is an Indian Cobra in its defense position. It is not "charmed" but is reacting to the presence of a possible aggressor by a threat display—raising the forepart of its body and drawing up its long anterior ribs to spread the skin of its neck into a hood.

Many of the elapids are unaggressive and seem loath to bite, but their

FIG. 16-12. The defense attitude of the South American Coral Snake, *Micrurus frontalis.*

poison is highly toxic. Some of the Cobras of Africa and Asia have the extremely unpleasant trait of "spitting" their venom into the eyes of their enemies. Their fangs are modified to permit the streams of venom to be ejected outward instead of downward. Spitting cobras can spray their poison with great force for distances of 180 cm. and they seem to show a high degree of accuracy in aiming for the eyes. When the poison is washed out immediately no permanent damage results, but untreated animals go blind.

The cobras (*Naja naja*) of the Kashmir and Punjab districts of India mate during January and February and their eggs are laid in May. Apparently the pair remain together from the time of mating until the young are hatched, and the male may also share in guarding the eggs. Incubation takes between 69 and 84 days. The usual number of eggs is rather low, from one to two dozen, but as many as 45 have been recorded.

Most elapids lay eggs, but the Ringhals or Spitting Cobras (*Hemachatus*) of South Africa and some of the Australian forms are ovoviviparous and *Denisonia* of Australia has been reported to be truly viviparous.

FAMILY HYDROPHIDAE

These are the true sea snakes. Since the largest only reach a length of 275 cm. and most species are only about one-third as long, they hardly qualify as the huge sea serpents of legend, but they may provide grounds for many such tales. They differ from the elapids mainly in the characters by which they are adapted to life in the sea. The body is more or less laterally compressed posteriorly and the tail is strongly compressed and paddle-shaped. Except in one genus (*Laticauda*) the nostril opens on the upper side of the snout and can be closed tightly by a valve. The pupil is round. The tongue is short so that only the cleft portion can be protruded. The sea snakes closely resemble the elapids in other ways and are often grouped with them as the proteroglyphous snakes because the grooved poison fangs are on the front of the maxillary bones.

The sea snakes are divided into two subfamilies.

Subfamily Laticaudinae

The three genera (*Laticauda, Aepysurus,* and *Emydocephalus*) included here are less specialized for marine life than are the members of the second subfamily. Their ventral shields are relatively large, being one-third to one-half as wide as the body. They are able to move around on land and apparently spend a good bit of time out of the water. At least two of the species, and possibly all of them, are oviparous, coming to shore to lay their eggs. They are never found far from land, but live in the shallow coastal waters and river estuaries of southeastern Asia, Australia, and the islands of Oceania. As would be expected, they feed on fish and are often taken in the nets of fishermen.

Subfamily Hydrophinae

These are the most aquatic of all the snakes. Their ventral scales are very small or absent (except those of *Ephalophis*) and, though graceful and accomplished swimmers, they are slow and awkward on land. Like the laticaudines, they are seldom found many miles out from shore, seeming to prefer the vicinity of coasts particularly around the mouths of rivers, where the waters are comparatively sheltered. Some species bask on the surface of the water and on days when the sea is calm, they may be seen from the bows of steamers, sometimes by the hundreds. One naturalist has described a mass of snakes forming a line across the surface of the sea about 300 cm. wide and nearly 100 km. long. The snakes were so closely packed together that the line could be seen from several miles away. He estimated that the column included millions of individual snakes—probably the largest congregation ever reported.

So far as is known, all the Hydrophinae are ovoviviparous. Although they are so aquatic, at least some species come to shore to bear their young. Female sea snakes of the Philippines have reportedly come up on the smaller islets to bring forth their young among the rocks and tidal pools.

About a dozen genera are included in this subfamily. They are found in the Indian and Pacific Oceans, along the Asiatic coast and throughout the Indo-Australian seas to the coast of tropical Australia and the oceanic islands of the southern Pacific. One form (*Pelamis*) has extended its range across the Pacific to the shores of tropical America and westward to Madagascar and Africa.

All of the sea snakes are poisonous. The venom of some does not appear to be strongly toxic to humans, but laboratory experiments have shown that the poison of others is even more powerful than that of the cobra. There are no records of bathers being attacked by them, and in general they can be induced to bite only after considerable provocation.

FAMILY VIPERIDAE

These are the snakes in which the whole mechanism for the injection of poison reaches its highest development. The maxillary bone is very short but deep vertically and is movably attached to the prefrontal and ectoptery-

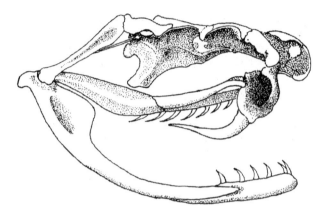

FIG. 16-13. The skull of the Florida Diamondback Rattlesnake, *Crotalus adamanteus.*

goid bones. The large poison fang is on its posterior end but because the bone is so shortened anteroposteriorly the fang lies in the front part of the mouth. The canal for the transmission of venom is usually completely closed so that no external groove is visible (solenoglyphous). At rest, the tooth is folded back to lie horizontally along the upper jaw. In striking, the fang is brought forward from the resting position by a movement of the bones forming the palatomaxillary arch, with the maxilla turning like a hinge on the anterior end of the prefrontal. Most elapids, with their smaller fixed fangs, tend to bite and hold on, but the viperids, with their large and powerful fangs, are able to inject a greater amount of poison at the instant of bite and they tend to strike and then draw back. Replacement of the fangs is the same as in the Elapidae, and two fangs are often present at the same time. There are no other maxillary teeth. In other characters, the viperids resemble the elapids.

Snake venoms are highly complex protein mixtures which vary in composition, and hence in effect, from species to species. In general, though, the poison of the elapids and hydrophids acts primarily on the nervous system (neurotoxic) while that of viperids acts on the blood (hemotoxic).

The Viperidae are divided into two subfamilies: the true vipers (Viperinae), and the pit vipers (Crotalinae).

Subfamily Viperinae

The true vipers are usually stockily built, with short bodies and short tails. The large Gaboon Viper (*Bitis gabonica*) may be nearly 180 cm.

long, but no viperid matches the big cobras in length. The maxillary bone is not hollowed out and there is no pit on the side of the face between the nostril and the eye. The 10 genera of true vipers are found only in Eurasia and Africa. Some are among the most deadly snakes in the world.

A few of the vipers lay eggs, but in most the young are born alive. A chorio-allantoic placenta forms in the European adder (*Vipera berus*). Mating takes place in the spring some time after the snakes have emerged from hibernation. Rivalry between males is keen during the mating season, and they sometimes put on a spectacular performance called the dance of the adders (see p. 114). Although copulation may take place in April, ovulation does not occur until about the end of May. The young are born during August and September. The eggs which will produce the young of the next year are formed in the ovaries following parturition. In the northern part of the range, the short summer is apparently insufficient for the complete intraovarian development of the eggs and the snake is forced back into hibernation before they are ready. In consequence, the vipers in the northern half of Sweden and in Finland breed only every second year. Numbers of the young range from 6 to 20, but there are usually about 10 to 14. They may be 130 to 180 mm. long.

FIG. 16-14. An Old World viperid, *Bitis gabonica*, the Gaboon Viper of Africa.

FIG. 16-15. A New World Pit Viper, *Lachesis muta stenophrys*, the Bush-master of South America.

Subfamily Crotalinae

These are vipers in which the maxillary bone is hollowed out above by a pit opening between the eye and nostril. The membrane in this pit is extremely sensitive to changes in temperature and serves to detect the presence of the warm-blooded animals on which the snake preys. Pit vipers occur from eastern Europe across Asia to Japan and the Indo-Australian archipelago, but they are most numerous in North, Central, and South America. There are five genera, including such dreaded forms as the rattle-snakes (*Crotalus*), and the tropical American Bushmasters (*Lachesis*) and Fer-de-Lance (*Trimeresurus*). The bushmaster reaches a length of 360 cm. and is truly one of the most formidable snakes in the world.

Apparently most of the crotalids except *Lachesis* are ovoviviparous. As in the viperids, there is an extensive premating dance of the males. The Cot-tonmouth Moccasin (*Agkistrodon piscivorus*) mates in March and the young are born in late August and early September. The number in a brood varies from 5 to 15 with an average of about 8. Length of the young ranges from 150 to more than 250 mm.

Collateral Reading and General Reference

Angel, F. *Vie et Moeurs des Serpentes*. Paris: Payot, 1950.

Bellairs, A. d'A. and G. Underwood. "Origin of Snakes." *Biological Reviews*, vol. 26, 1951. (A fairly recent paper that has already become a classic.)

Boulenger, G. A. *Catalogue of the Snakes in the British Museum* (*Natural History*). Vols. 1–3, London: British Museum, 1893–1896.

Cope, E. D. *The Crocodilians, Lizards, and Snakes of North America*. Report of the United States National Museum for 1898, pt. 2, 1900.

Klauber, L. M. *Rattlesnakes*. 2 vols. Berkeley and Los Angeles: University of California Press, 1956. (Probably the finest monograph ever published on a single group of herptiles.)

Pope, C. H. *Snakes Alive*. New York: Viking Press, 1942. (A good, sound introduction to the biology of snakes.)

Smith, M. A. *Fauna of British India. Reptilia and Amphibia*. Vol. 3. *Serpentes*. London: Taylor and Francis, 1943.

Wright, A. H. and A. A. Wright. *Handbook of Snakes*. Vols. 1 and 2. Ithaca: Comstock, 1957. (Contains a wealth of natural history information, though limited geographically to the snakes of the United States and Canada.)

RHYNCHOCEPHALIANS

AND CROCODILIANS

THE TWO REMAINING ORDERS of reptiles are both relics. The Tuatara (*Sphenodon punctatus*) is the only surviving member of the order Rhynchocephalia, now placed in the subclass Lepidosauria to which the Squamata also belong. No fossil remains of *Sphenodon* have ever been found, but all other members of the family are known only from the Triassic and Jurassic. In spite of this enormous time gap, *Sphenodon* seems to have changed little down through the ages and remains today as a relatively unspecialized representative of the reptiles of the early Mesozoic. It has been aptly called a "living fossil." The modern crocodilians are all that remain of the mighty Archosaur stock that once throve and gave rise to the Mesozoic dinosaurs, as well as to the modern birds.

Although they are placed in different subclasses, in one respect *Sphenodon* and the crocodilians resemble each other and differ from all other living reptiles. They have diapsid skulls, in which both dorsal and temporal fossae, with their bounding arches, are present. The turtles have anapsid skulls, without fossae, and the lizards and snakes have lost one or both of the arches.

ORDER RHYNCHOCEPHALIA

These primitive reptiles look like lizards, but differ from them not only in having a two-arched skull but in many other structural characters. Teeth are present on the premaxillary, maxillary, palatine, and dentary and vestigially on the vomer. Well-developed gastralia are present. The male lacks

312

a copulatory organ. The cloacal opening is a transverse slit. A nictitating membrane, or third eyelid, can be moved slowly across the eyeball from the inner corner of the eye outward while the upper and lower lids remain open. A well-developed parietal eye, with small lens and retina, is present on top of the head. In the young it can be seen clearly through the translucent covering scale, but in the adult the skin above thickens. A similar structure is present in many lizards. It may be sensitive to heat and light.

Family Sphenodontidae

Sphenodon punctatus is scarcely known to most nonzoologists, but to the professional it is one of the most fascinating creatures alive. It shows us, in the flesh, what some of the early reptiles of the Mesozoic must have been like. One of the most perplexing things about it is its apparent failure either to evolve or to become extinct. If we could learn why it has remained virtually unchanged through such long ages of time, we might understand more of the forces that do bring about evolutionary change in most living things.

The adult Tuatara is about 500 to 800 mm. long. It is brownish-olive, and has a small yellow spot in the center of each scale. Enlarged, spiny scales form a crest down the back and tail.

Tuataras are found only in New Zealand. Those on the main islands succumbed rapidly to the onslaught of the mammals introduced by European settlers, and the remainder now inhabit only the waterless offshore islands, where they are rigidly protected. They live in close association with vast colonies of nesting shearwaters, called Mutton Birds by the New Zealanders. These birds nest in underground burrows, which they share with the Tuataras. For the most part the association seems amicable, although the normally insectivorous Tuataras occasionally feed on eggs or nestling birds.

The Tuatara remains in its burrow during the day, coming out to prowl at night when the temperature drops sharply and cold gusts of wind sweep over the islands. These animals are active at lower temperatures than other reptiles, and their body temperature tends to be lower than that of their surroundings. Body temperatures ranging from 6.2 to 13.3 C. have been reported in nature.

Mating has not been observed, but since the male lacks a copulatory organ, it is believed that insemination takes place by simple cloacal apposition. During the summer season, from November to January, the female lays about 10 white, hard-shelled, elongate eggs about 28 mm. in length. They are usually deposited well away from the home burrow in a shallow hole in sand where they can be warmed by the sun. By August, the embryos are nearly mature. However, the late stage embryo apparently undergoes a sort of aestivation over the second summer and does not hatch until about 13 months old. During this aestivation period, the nasal chambers become blocked with a proliferating epithelium which is resorbed shortly before hatching.

ORDER CROCODILIA

Crocodilians are elongated reptiles with a muscular, laterally compressed tail, a more or less elongated snout, and two pairs of short legs. There are five toes on the front feet, and four on the hind feet. A nictitating membrane is present. The cloacal opening is a longitudinal slit, the penis is single, and there is no urinary bladder. The tongue is not protrusible. Gastralia and dorsal and ventral epidermal scales, reinforced by bony plates, are present. The skull is diapsid.

In some respects, crocodilians are the most advanced of all living reptiles. They possess a true cerebral cortex and a completely four-chambered heart. The teeth are thecodont, i.e., set in sockets in the jaw. The skin of the head is fused to the skull bones and there are no fleshy lips to make a water-tight closure of the mouth possible. The nostrils are far forward on top of the elongate snout. The maxillaries, palatines, and pterygoids meet in the midline of the roof of the mouth to form a secondary palate which separates the nasal passages from the mouth. These air passages open into the throat behind a valve formed by a fleshy fold at the back of the tongue which meets a similar fold on the palate. Water in the mouth is thus kept separate from the inspired air and the crocodile is able to breathe while submerged with only the tip of its snout protruding or while holding prey in the water.

Although *Sphenodon* is of interest mainly to zoologists, crocodilians have a horrible fascination for almost everyone. They truly look like monsters out of the prehistoric past. The huge size attained by some—the maximum length recorded for an American Crocodile is 690 cm.—and the man-eating proclivities of a few, so color our concept of the whole group that we think of all crocodiles as ferocious monsters. Actually, some are dwarf forms quite harmless to man. The Congo Dwarf Crocodile has a maximum known length of 1125 mm.

Evolution in the crocodiles has progressed along two lines. Some forms have a relatively short and broad snout, and the two halves of the lower jaw are joined by a short symphysis. Others have a long, narrow snout, and the symphysis of the lower jaw is long. We recognize eight genera of crocodilians, seven in the family Crocodylidae and the eighth, *Gavialis*, in a separate family, Gavialidae.

Family Crocodylidae

The snout of the true crocodiles is not sharply set off from the posterior part of the skull. The maxillary bones do not meet dorsally to separate the

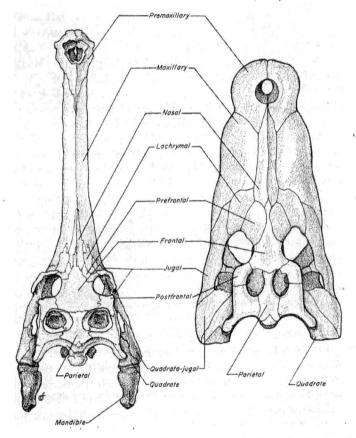

FIG. 17-1. The skull of *Gavialis gangeticus* (left) and of *Crocodylus palustris* (right) showing the disposition of the nasal bones (separated from the premaxillaries by the maxillaries in the former but not in the latter). [After Malcolm Smith.]

nasals from the premaxillaries. There are 14 to 24 teeth on each side of the lower jaw.

The genus *Crocodylus* is widely distributed, being found in the tropical and subtropical regions of North and South America, the West Indies, Africa, Asia, Australia, and the East Indies. It includes both broad-snouted forms, such as the Indian Mugger (*Crocodylus palustris*) and narrow-snouted ones like the African *Crocodylus cataphractus*, in which the mandibular symphysis extends back to the level of the eighth tooth. The genus *Osteolaemus* comprises two dwarf species found in Africa. *Tomistoma*, the False Gavial of Borneo, Sumatra, and the Malay Peninsula, is the most narrow-snouted of all the Crocodylidae, with the mandibular symphysis extending back to the fourteenth or fifteenth tooth.

The broad-snouted, predominantly New World genera are sometimes placed in a separate subfamily, or even family, Alligatoridae, but since *Crocodylus* also includes broad-snouted forms, and since the distinction between the two groups breaks down when fossil forms are considered, it seems best to include them all in the Crocodylidae. *Alligator* is the most northern in distribution of all the crocodilians, being found in the southeastern United States and in southern China. Three genera of alligator-like Caimans are found in Central and South America. These are *Caiman*, *Melanosuchus*, and *Paleosuchus*.

All crocodilians are aquatic, living in rivers, marshes, and lakes, and some take freely to salt water. They frequently bask in the sun on shore but most species seldom wander far from water. The long, strong tail is used for swimming and is also a powerful weapon of offense or defense. By it a victim can be literally swept off his feet and is then seized and dragged into the water before he can recover and escape. If prey is too large to be swallowed whole the crocodile tears it limb from limb by the gruesomely effective methods of rotating its whole body rapidly over and over in the water.

The sharp pointed teeth are used for seizing but not for chewing. The muscular stomach functions much as does the gizzard of a bird and, like a bird, a crocodilian swallows hard objects to aid in grinding up the stomach contents. Sixty-nine pebbles were found in the stomach of a Smooth Fronted Caiman (*Paleosuchus*). In the southeastern United States, where stones are scarce, alligator stomachs have been found to contain Coca-cola bottles, bottle tops, the brass portions of shot gun shells, and "lightered knots" as the natives call hard knots of resinous pine.

Variation in the number and position of the bony plates that underlie the horny scales in the different species affects the usefulness of the hide. The

FIG. 17-2. Young Spectacled Caimans, *Caiman sclerops*, of the Amazon and Orinoco Rivers of South America.

American Alligator yields a very sturdy and beautiful leather and has been widely hunted. In 1937, 135,000 hides were bought by dealers in Florida at $4.00 a hide. By 1943 the price had risen to $19.75 for a 7-foot hide, but excessive hunting had so reduced the population that only 6,800 were taken. Florida then passed a law protecting the alligator during the breeding season and prohibiting the capture of specimens less than 4 feet long. By 1947 the take had risen to 25,000 hides, and the price had dropped to $13.30 a hide.

A few lizards and turtles have voices, but for the most part reptiles are silent creatures. Crocodilians are an exception. Young of some species make a curious, high-pitched croak when disturbed. The voice of the adult male is something between a deep bark and a bellow. In the spring months the bellowing of a bull alligator is perhaps the most impressive sound of the southeastern swamps, and is something that every herpetologist looks forward to hearing.

All crocodilians lay large, hard-shelled eggs, either in a shallow excavation on a sandy shore or in a nest piled up by the mother. The female American Alligator scoops up mud and vegetation with her jaws to build a mound about 90 cm. high and 150 to 210 cm. wide at the base. She deposits 20 to 70 eggs in a hollow in the center and covers them over with material from the rim of the nest. She remains on guard nearby for 9 or 10 weeks until she hears the young, now ready to emerge, being to peep loudly. She then tears open the nest and allows them to escape.

Family Gavialidae

The gavial has an extremely long and slender snout which is sharply set off from the posterior part of the skull. There are 25 to 26 teeth on each side of the lower jaw and the mandibular symphysis extends back to the level of the 23rd or 24th tooth. The maxillary bones meet dorsally to separate the nasals from the premaxillaries. The maximum recorded length of the gavial is 645 cm.

The single living species, *Gavialis gangeticus*, is found only in India and Burma. Its colloquial name in India is gharial. The generic name was based on this, but through a clerical error was published as *Gavialis* rather than *Garialis*. Now the misformed scientific name has given rise to a new common name, gavial, which has largely replaced the original gharial.

The very narrow snout is an adaptation for eating fish, which the gavial catches by a sudden, sidewise sweep of the head through the water. A broad snout could not be moved so rapidly. The gavial has been reported to catch birds and such fair-sized mammals as goats and dogs occasionally, but in spite of its large size it seldom if ever attacks man.

The female lays 40 or more eggs in a hole in a sand bank. These eggs are 85–90 mm. long by 65–70 mm. wide. The young, which appear in March and April, are about 375 mm. long. Obviously they must have been very tightly coiled up within the eggs.

Collateral Reading and General Reference

Cope, E. D. *The Crocodilians, Lizards, and Snakes of North America*. Report of the United States National Museum for 1898, pt. 2, 1900.

McIlhenny, E. A. *The Alligator's Life History*. Boston: Christopher Publishing House, 1935. (Based on many years of personal experience with these great reptiles.)

Mertens, R. and H. Wermuth. *Die rezenten Schildkröten, Krokodile und Brükenechsen*. Zoologische Jahrbücher, Band 83, Heft 5, 1955.

Reese, A. M. *The Alligator and its Allies*. New York: Putnam's, 1915.

Smith, M. A. *Fauna of British India. Reptilia and Amphibia*. Vol. I. *Loricata, Testudines*. London: Taylor and Francis, 1931.

CLASSIFICATION

The classification of amphibians and reptiles followed in this book is outlined below. All groups are classified to the ordinal level, and those orders, or suborders, having living representatives, to the level of families. Groups that are entirely fossil are indicated by an asterisk.

CLASS AMPHIBIA

*Superorder Temnospondyli
  *Order Ichthyostega
  *Order Rhachitomi
  *Order Trematosauria
  *Order Stereospondyli
Superorder Lepospondyli
  *Order Aistopoda
  *Order Nectridia
  Order Trachystomata
    Family Sirenidae
  *Order Microsauria
  Order Apoda
    Family Caecilidae
  Order Caudata
    Suborder Cryptobranchoidea
      Family Hynobiidae
      Family Cryptobranchidae
    Suborder Ambystomatoidea
      Family Ambystomatidae
    Suborder Salamandroidea
      Family Salamandridae
      Family Amphiumidae
      Family Plethodontidae
    Suborder Proteida
      Family Proteidae
    Caudata *Incertae Sedis*
      *Family Batrachosauroididae
      *Family Scapherpetonidae

Superorder Salientia
  Order Anura
    Suborder Amphicoela
      *Family Notobatrachidae
      Family Leiopelmidae
    Suborder Aglossa
      Family Pipidae
      *Family Palaeobatrachidae
    Suborder Opisthocoela
      Family Discoglossidae
      Family Rhinophrynidae
      *Family Montsechobatrachidae
    Suborder Anomocoela
      Family Pelobatidae
      Family Pelodytidae
    Suborder Diplasiocoela
      Family Ranidae
      Family Rhacophoridae
      Family Microhylidae
      Family Phrynomeridae
    Suborder Procoela
      Family Pseudidae
      Family Bufonidae
      Family Atelopodidae
      Family Hylidae
      Family Leptodactylidae
      Family Centrolenidae

°Superorder Anthracosauria   °Order Seymouriamorpha
  °Order Embolomeri

CLASS REPTILIA

Subclass Anapsida
  °Order Cotylosauria
  Order Testudinata
    °Suborder Amphichelydia
      (several fossil families)
    Suborder Pleurodira
      Family Pelomedusidae
      Family Chelidae
    Suborder Cryptodira
      Superfamily Testudinoidea
        Family Dermatemydidae
        Family Chelydridae
        Family Testudinidae
      Superfamily Chelonioidea
        °Family Toxochelyidae
        °Family Protostegidae
        °Family Desmatochelyi-
          dae
        Family Cheloniidae
      Superfamily Dermochelyoi-
        dea
        Family Dermochelyidae
      Superfamily Carettochelyoi-
        dea
        Family Carettochelyidae
      Superfamily Trionychoidea
        Family Trionychidae
Subclass Lepidosauria
  °Order Eosuchia
  Order Rhynchocephalia
    Family Sphenodontidae
    °Family Rhynchosauridae
    °Family Sapheosauridae
    °Family Claraziidae
    °Family Pleurosauridae
  Order Squamata
    Suborder Lacertilia
      Infraorder Gekkota
        Family Eublepharidae
        Family Sphaerodactyli-
          dae
        Family Gekkonidae

        Family Pygopodidae
        Family Dibamidae
      Infraorder Iguania
        Family Iguanidae
        Family Agamidae
      Infraorder Rhiptoglossa
        Family Chamaeleonidae
      Infraorder Scincomorpha
        °Family Ardeosauridae
        Family Scincidae
        Family Anelytropsidae
        Family Feylinidae
        Family Cordylidae
        Family Lacertidae
        Family Teiidae
      Infraorder Anguinomorpha
        Superfamily Diploglossa
          Family Anguinidae
          Family Aniellidae
          Family Xenosauridae
        Superfamily Platynota
          Family Helodermati-
            dae
          Family Varanidae
          Family Lanthonotidae
          °Family Aigialosauridae
          °Family Mososauridae
          °Family Dolichosauri-
            dae
          °Family Palaeophidae
      Infraorder Annulata
        Family Amphisbaenidae
      Lacertilia Incertae Sedis
        Family Xantusiidae
        Family Typhlopidae
    Suborder Serpentes
      °Family Dinilysiidae
      Family Boidae
      Family Aniliidae
      Family Uropeltidae
      Family Xenopeltidae
      Family Leptotyphlopidae

*Family Archaeophiidae
Family Acrochordidae
Family Colubridae
Family Elapidae
· Family Hydrophidae
Family Viperidae
Subclass Archosauria
*Order Thecodontia
Order Crocodilia
*Suborder Protosuchia
(one fossil family)
*Suborder Mesosuchia
(six fossil families)
Suborder Eusuchia
*Family Stomatosuchidae
Family Crocodylidae
Family Gavialidae

*Suborder Sebacosuchia
(two fossil families)
*Suborder Thalattosuchia
(one fossil family)
*Order Pterosauria
*Order Saurischia
*Order Ornithischia
*Subclass Ichthyopterygia
*Order Ichthyosauria
*Subclass Euryapsida
*Order Protorosauria
*Order Sauropterygia
*Subclass Synapsida
*Order Pelycosauria
*Order Therapsida
Reptilia *Incertae Sedis*
*Order Mesosauria

AMPHIBIAN AND REPTILIAN CHROMOSOMES

The following table gives the chromosome number, number of metacentric and acrocentric chromosomes, and N. F. (Nombre Fondamental) for those species of amphibians and reptiles for which this information has been recorded. Chromosome counts have been made on a number of other species of herptiles, but the N. F.'s have not been worked out. These other species are listed in: Makino, S., *Chromosome Numbers in Animals.*, 2d Ed., Iowa State College Press, 1951.

Family	Scientific Name	2N	Meta-centric	Acro-centric	N. F.
	AMPHIBIA				
Caecilidae	*Ichthyophis glutinosus*	42	10	32	52
	Uraeotyphus narayani	36	16	20	52
Hynobiidae	*Hynobius kimurai*	60	16	44	76
	Hynobius lichenatus	58	18	40	76
	Hynobius retardatus	40	22	18	62
	Hynobius keyserlingi	62	12	50	74
	Ten other species of *Hynobius*	56	20	36	76
	Pachypalaminus boulengeri	56	20	36	76
Cryptobranchidae	*Cryptobranchus alleganiensis*	62	12	50	74
	Megalobatrachus japonicus	64	10	54	74
Ambystomatidae	*Ambystoma mexicanum*	28	28	0	56
	Ambystoma maculatum	28	28	0	56
Salamandridae	*Pleurodeles waltli*	24	24	0	48
	Triturus cristatus	24	24	0	48
	Triturus vulgaris	24	24	0	48
	Triturus alpestris	24	24	0	48
	Triturus helveticus	24	24	0	48
	Salamandra salamandra	24	24	0	48
	Ten other species of *Salamandridae*	24	24	0	48
Plethodontidae	*Desmognathus fuscus*	28	28	0	56
Proteidae	*Proteus anguinus*	18	18	0	36
	Necturus maculosus	24	24	0	48

Family	Scientific Name	2N	Meta-centric	Acro-centric	N. F.
Amphiumidae	*Amphiuma means*	24	24	0	48
Leiopelmidae	*Ascaphus truei*	42	12	30	54
Pipidae	*Pipa pipa*	22	8	14	30
	Xenopus laevis	36	12	24	48
Discoglossidae	*Discoglossus pictus*	28	20	8	48
	Alytes obstetricans	36	12	24	48
	Bombina bombina	24	24	0	48
	Bombina orientalis	24	24	0	48
	Bombina variegata	24	24	0	48
Pelobatidae	*Pelobates fuscus*	26	26	0	52
Ranidae	*Rana arvalis*	26	26	0	52
	Rana catesbeiana	26	26	0	52
	Rana dalmatina	26	26	0	52
	Rana esculenta	26	26	0	52
	Rana graeca	26	26	0	52
	Rana limnocharis	26	26	0	52
	Rana nigromaculata	26	26	0	52
	Rana palustris	26	26	0	52
	Rana pipiens	26	26	0	52
	Rana rugosa	26	26	0	52
	Rana temporaria	26	26	0	52
Rhacophoridae	*Rhacophorus schlegeli*	26	26	0	52
	Rhacophorus buergeri	26	26	0	52
Microhylidae	*Kaloula borealis*	28	28	0	56
Bufonidae	*Bufo americanus*	22	22	0	44
	Bufo arenarium	22	22	0	44
	Bufo bufo	22	22	0	44
	Bufo calamita	22	22	0	44
	Bufo canorus	22	22	0	44
	Bufo fowleri	22	22	0	44
	Bufo terrestris	22	22	0	44
	Bufo quercicus	22	22	0	44
	Bufo raddei	22	22	0	44
	Bufo regularis	22	22	0	44
	Bufo sachalinensis	22	22	0	44
	Bufo viridis	22	22	0	44
Hylidae	*Acris gryllus crepitans*	22	22	0	44
	Acris g. gryllus	22	22	0	44
	Hyla arborea	24	24	0	48
	Hyla avivoca	24	24	0	48
	Hyla cinerea	24	24	0	48
	Hyla versicolor	24	24	0	48
	REPTILIA				
Chelydridae	*Sternotherus odoratus*	50	4	46	54
Testudinidae	*Emys orbicularis*	50	4	46	54

Family	Scientific Name	2N	Meta-centric	Acro-centric	N. F.
	Clemmys japonica	52	8	44	60
Trionychidae	*Trionyx japonica*	64	6	58	70
Sphenodontidae	*Sphenodon punctatus*	36	12	24	48
Gekkonidae	*Tarentola mauritanica*	42	0	42	42
	Hemidactylus bowringi	46	0	46	46
	Hemidactylus flaviridis	46	0	46	46
	Gekko japonicus	38	4	34	42
	Gymnodactylus milliusi	38	0	38	38
	Eublepharis variegatus	32	0	32	32
Iguanidae	*Anolis carolinensis*	36	12	24	48
Agamidae	*Agama stellio*	36	12	24	48
	Uromastix hardwicki	36	12	24	48
	*Japalura swinhonis	46	0	46	46
	Calotes versicolor	34	12	22	46
	Sitana ponticeriana	46	0	46	46
Chamaeleonidae	*Chamaeleo chamaeleon chamaeleon*	24	24	0	48
Scincidae	*Scincus officinalis*	32	4	28	36
	Chalcides chalcides	28	8	20	36
	Eumeces latiscutatus	26	12	14	38
	Mabuya macularia	26	10	16	36
Cordylidae	*Gerrhosaurus flavigularis*	36	12	24	48
Lacertidae	*Lacerta viridis*	38	0	38	38
	Lacerta agilis	38	0	38	38
	Lacerta muralis	38	0	38	38
	Lacerta vivipara	36	0	36	36
	Lacerta ocellata	36	2	34	38
	Tachydromus tachydromoides	38	0	38	38
	Tachydromus formosanus	38	0	38	38
	Tachydromus septentrionalis	38	0	38	38
	Psammodromus algirus	38	0	38	38
	Psammodromus hispanicus	38	0	38	38
Teiidae	*Tupinambis teguixin*	36	10	26	46
	Cnemidophorus sexlineatus	46	2	44	48
	Ameiva surinamensis	50	0	50	50
Varanidae	*Varanus gouldi*	40	8	32	48
Helodermatidae	*Heloderma suspectum*	38	10	28	48
Anguinidae	*Zonurus cataphractus*	46	0	46	46
	Anguis fragilis	44	4	40	48
	Ophisaurus apodus	44	4	40	48
	Ophisaurus ventralis	30	4	26	34
	Gerrhonotus kingi	45	1	44	46
	Gerrhonotus multicarinata scincicauda	40	4	36	44
	or	42	2	40	44
Amphisbaenidae	*Trogonophis wiegmanni*	36	12	24	48

Family	Scientific Name	2N	Meta-centric	Acro-centric	N. F.
	Rhineura floridana	46	2	44	48
Xantusiidae	*Xantusia henshawi*	42	6	36	48
Colubridae	*Natrix natrix*	36	10	26	46
	Natrix maura	36	10	26	46
	Natrix tigrina	40	10	30	50
	Dinodon rufozonatum	46	2	44	48
	Macropistodon rudis	46	2	44	48
	Oligodon formosanus	36	10	26	46
	Zaocys nigromarginatus	36	10	26	46
	Elaphe quadrivirgata	36	10	26	46
	Elaphe climacophora	36	10	26	46
	Coluber gemonensis	36	10	26	46
	Coronella austriaca	36	10	26	46
	Telescopus fallax	36	10	26	46
	Coelopeltis lacertina	42	6	36	48
Elapidae	*Bungarus multicinctus*	36	4	32	40
	Naja naja	38	10	28	48
Hydrophidae	*Laticauda semifasciata*	38	8	30	46
Viperidae	*Ancistrodon acutus*	36	10	26	46
	Ancistrodon halys	36	10	26	46
	Trimeresurus gramineus	36	10	26	46
	Trimeresurus mucrosquamatus	36	10	26	46
	Vipera aspis	42	4	38	46
	Vipera berus	36	10	26	46
Crocodylidae	*Alligator mississippiensis*	32	10	22	42

* It has been found that the N. F. of a number of the lizards is lower than might be expected judging by other related genera. Matthey (1949) suggests that this is due to fusions of micro-acrocentric chromosomes to macro-acrocentrics, or in some cases to the loss of micro-chromosomes. He believes that the basic N. F. of all Lacertilians except the Gekkonidae is in reality about 48 and gives a corrected N. F. where such fusions or losses, or both, seem to have occurred. The N. F.'s given here have not been so corrected.

A number in italics indicates the page on which the main discussion of an item appears. A number in boldface indicates a page on which an item is illustrated.

327

F

red
orange
yellow

green

blue
indigo
violet

CPSIA information can be obtained
at www.ICGtesting.com
Printed in the USA
BVHW060008040522
635994BV00034B/942